EMERGENCY EXERCISE HANDBOOK

Evaluate and Integrate Your Company's Plan

EMERGENCY EXERCISE HANDBOOK

EVALUATE AND INTEGRATE YOUR COMPANY'S PLAN

By

Tracy Knippenburg Gillis

PennWell Books

Pennwell Publishing Company

Tulsa, Oklahoma

PennWell Publishing Company
1421 South Sheridan/P.O. Box 1260
Tulsa, Oklahoma 74101

Library of Congress Cataloging-in-Publication Data

Printed in the United States of America

1 2 3 4 5 00 99 98 97 96

Contents

Foreword

The author and her firm, The Corporate Response Group, Inc., a worldwide crisis management consulting firm, are crisis management experts specializing in part in preparing and conducting exercise activities for clients in the private and public sectors. Their experiences are broad, crossing industries and country boundaries. As such, the author and contributors have written this book based on a wealth of experience in what does and does not work well in the real world. The book attempts to share the benefits of these experiences and to accommodate the diverse organizations, cultures, requirements, and expectations of its readers.

Acknowledgements

Much like an exercise, this book is not the work of just one person. I should first acknowledge the contributions of hundreds of people who have asked "How?" or "Why?" Through these questions emerged the realization that there exists a need for information and a systematic process to exercising. I went on to assemble the guidance included in this book based on those many consultations.

I must also thank my staff and peers at the Corporate Response Group who helped me with this project. Especially among those I would like to thank are Terry Wells, Chandra Gribbon, and Bob Wilkerson for early versions of this material, as well as Tom E. Davis for his early support and guidance on publishing. Then there are Lisa Hawes and Nichole Shumanis, who were critical in supporting completion of the final text of this book. Special thanks go to Alfredo Suarez, for his graphic design. And there are certainly many more who were supportive in so many ways.

Finally, but certainly not last in my life and in the writing of this book, I express my appreciation to my partner in life, John. He has been supportive throughout and has been more understanding than any wife deserves.

Introduction

Ours is a world of ever-changing risks and vulnerabilities. In the high-tech, advanced society we live in, there are inherent hazards we as communities, businesses, and citizens face on a daily basis. Businesses are expected to be prepared to respond to protect life, property, and the environment, in addition to preserving stakeholder value and the company image. In order to accomplish this effectively and efficiently, sound crisis and emergency management capabilities must be in place. This book describes the best process of validating—in fact, proving—that your organization's capabilities are what they should be and are adequate to meet the worst-case situations that may occur. Short of a real event, exercising is the only way to be sure you can respond and manage an emergency or crisis and the associated consequences.

The purpose of this book is

- to increase your knowledge of exercising and how to create an exercise program that will maximize the value of the resources you are devoting to crisis and emergency management,
- to provide you with a methodology and process for creating high-quality exercise activities,
- to improve your understanding and ability to link your exercise efforts to your broader crisis management program.

To accomplish this, we have designed this book to promote learning through narrative content presentation, presentation of samples, and activities that allow you to practice the methodologies and techniques described in each chapter. The sound process and methodology provided in this book are designed to support any organization, whether within the public or private sector. The process will work regardless of the industry in which you work, the country in which you operate, or the level in the organization you represent.

As you read the book, you will need to make a broad assessment of your organization's crisis and emergency management capability in order to take a programmatic approach to exercising. The book will guide you step-by-step through a process that allows you to

- assess your organization's crisis and emergency management capabilities,
- create a multi-year exercise program to accomplish your organization's exercise objectives,
- discover how to design an exercise activity for your organization,
- learn the process of developing exercise activities,

- recognize the importance and learn the techniques of effective exercise conduct,
- enhance your organization's exercise conduct skills specifically in the areas of control, evaluation, and simulation,
- become familiar with the range of evaluation methodologies and available tools and techniques,
- learn about tracking exercise results to maximize results and ensure sustained improvements.

The book will begin by defining basic exercise terminology and reviewing the philosophy and history of exercising.

CHAPTER 1

Exercising Basics

Contrary to popular belief, crises do not bring out the best in our leaders, but in fact, often produce flawed decision-making. Senior officials are often too busy to participate in crisis simulations, yet when crises occur–when the unimaginable becomes reality–senior officials will inevitably be involved.

Jerrold Post, M.D.
Director of Political Psychology
Professor of International Affairs
George Washington University

This chapter is designed to provide a basic understanding of exercising. It describes the process of establishing a comprehensive exercise program and designing, developing, conducting, and evaluating valuable, high-quality exercise activities.

The objectives of this chapter are to help you

- recognize the history and philosophy of exercising,
- increase your knowledge of the different types of exercises and the associated terminology,
- understand when to use each exercise most effectively.

Introduction

As we begin this exercising book, it is important to review some of the basics of exercising. This will help ensure the proper focus and a common understanding and use of terminology. It will also provide the foundation for building an effective exercise program and developing high-quality exercise activities.

Why Exercise?

Exercising provides a means for organizations to validate their capability without actually going through real-life incidents. The benefits of this type of activity for the organization are numerous. While adding value to the organization, exercising also meets the needs and requirements of external parties. There are both internal and external benefits to exercising.

The internal benefits of exercising include

- enhancing capability,
- improving coordination and proficiency,
- finding areas in need of program improvement,
- identifying resource requirements,
- validating existing plans and procedures.

The externally driven benefits of exercising include

- meeting regulatory requirements,
- responding to political and public pressure,
- demonstrating a commitment to preparedness and meeting expectations.

One way of viewing exercising would be to apply the old adage, "Practice makes perfect."

Philosophy of Exercising

Your approach to exercising and your organization's exercise program should be based upon several principles that represent the overall philosophy behind exercising. The following are the essential aspects of a sound philosophy.

Validate vs. Test

One of the fundamental principles of exercising is to *validate* capabilities. The term *validate* is preferred over the term *test* because *test* tends to imply a pass/fail relationship or the possibility of performing poorly. Exercise activities should measure performance in order to identify areas for improvement, rather than to indicate a failure to perform. This philosophy is increasingly being adopted by regulators, in addition to industry.

Training vs. Exercising

A second principle of exercising is that it should always provide positive learning opportunities and enable players to demonstrate the use of skills and knowledge. Successful exercises and simulations maximize learning opportunities by linking the lessons learned and areas for improvement throughout the organization. These linkages lead to a cycle of continuous improvement that extends beyond the exercise participants.

Mix of Exercise Activities

Exercise activities range from single-function, single-level, discussion-based activities, to more complex, multiple-function, action-based activities involving hundreds or even thousands of personnel. Utilizing a mix of exercise activities over time in your exercise program ensures an adequate level of capability is maintained, while ensuring the program is cost-effective.

Exercise Scheduling and Cycle

Exercising efforts should be scheduled and programmed into a cycle of activity. Not only should your exercise program rely on a mix of exercise activities, as described above, but it should build upon and interrelate to the other activities that are a part of your overall emergency and crisis management program. The following graphic in Figure 1-1 depicts the factors that support scheduling an appropriate cycle of exercise activities. The elements associated with scheduling are discussed in Chapter 3 of this book.

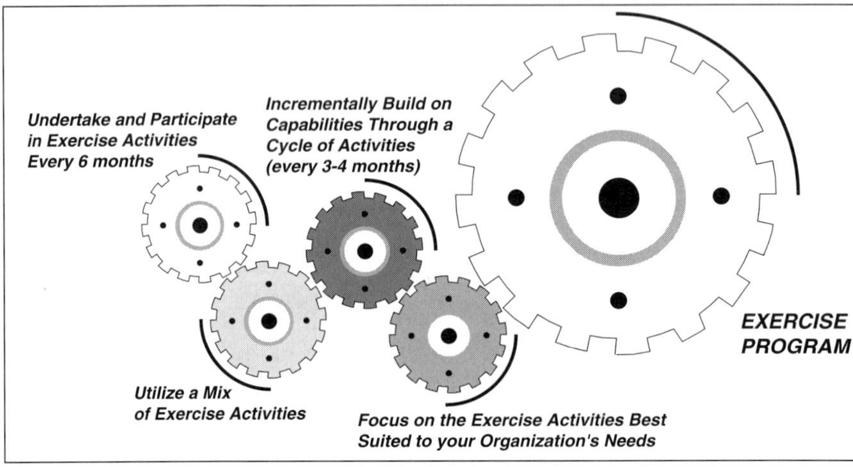

Figure 1-1
Exercise Activity Scheduling Guidelines

Exercising Definitions/Descriptions

In order to ensure a common foundation from which to discuss design, development, conduct, and evaluation, we must first have a common understanding of terminology. Definitions are provided for terms relevant to each of these important phases and your exercise program, as well as for events requiring response. It is important to remember that each industry, agency, and organization may define specific exercise activities to meet its unique purposes, requirements, or needs. Therefore, these definitions may not identically match those you may already be using, but our experience suggests that these are commonly agreeable definitions. Further, some of these terms may not yet be familiar to you. I ask that you embrace them and begin incorporating them into your lexicon as you see fit. Refer to the glossary for additional terms and definitions.

First, crisis and emergency management terms:

Incident: An event, series of events, or set of circumstances that interrupts normal operating procedures and has the potential to precipitate an emergency or crisis.

Emergency: An unforeseen combination of circumstances that disrupts normal operating conditions and poses a potential threat to human life, health, property, and/or the environment if not controlled, contained, or eliminated immediately.

Crisis: Any incident, series of events, or set of circumstances that threatens to fundamentally affect or alter the way an organization conducts business.

Disaster: The consequences of an incident, emergency, or crisis on the surrounding area. Typically, but not always, a disaster is associated with a natural disaster such as a tornado, hurricane, or earthquake.

Emergency Management: Those measures taken to prevent, prepare for, respond to, and recover from the acute effects of an emergency. The term is often used interchangeably with *Incident Response* or *Disaster Management*. This process/term is discussed further in Chapter 2.

Crisis Management: A business philosophy that advocates capabilities to prevent, prepare for, respond to, and recover from an incident or set of circumstances that has the potential to quickly cause radical changes in the way an organization conducts its business. This process/term is discussed further in Chapter 2.

Emergency/Crisis Management Program: A program that encompasses all of your organization's prevention, preparedness, response, and recovery efforts, including such aspects as planning, quality assessments, exercising, training, and communications.

Next, exercise activities:

Exercise Activity: An umbrella term describing an activity that provides an opportunity for players to train in and practice emergency and crisis management skills. Exercise activities provide a method to evaluate player's ability to meet emergency and crisis management requirements and responsibilities.

Tabletop Exercise: An exercise activity conducted in a conference room-type setting involving discussion *only* of a scenario by a response team or teams.

Drill: An exercise activity requiring a focused performance of only limited response activities or functions by individuals and teams, and limited mobilization of personnel and resources.

Exercise: An exercise activity requiring performance, integration, and coordination of response activities by several individuals and teams, as well as mobilization of personnel and resources.

Exercise Program: A combination of various exercise activities (tabletop exercises, limited and full drills and exercises) scheduled over a multi-year cycle to build and maintain individual, team, and organization capabilities. An Exercise Program is based upon objectives and is typically coordinated with the overall emergency or crisis management program activities.

Full-Scale Exercise: An exercise activity that validates the major aspects of an organization's crisis management program. It involves all levels of the organization.

Although these definitions do not support the philosophy, some organizations use the terms *drill* and *exercise* interchangeably. A number of other terms are used frequently when discussing exercise activities. The following provides an explanation of these terms.

Simulation: An umbrella term that means an artificially produced condition that replicates a real-life emergency or crisis situation. This can refer to any type of exercise activity or can also refer to the process of role-playing.

Command Post Exercise: This is industry-unique terminology, typically referring to a simulation that is a combination of a tabletop activity and a drill.

Walk-Through: This term is frequently used as an exercise activity term. However, our definition (and our belief) is that a *walk-through* is a training activity that typically guides responders through their response facilities, use of equipment, and the general concept of operations.

Finally, basic exercising terms:

Trusted Agent: Individuals with unique or specialized expertise who are confidentially included in the scenario development to ensure realistic events are postulated and appropriate responses are anticipated.

Talk-Through: Sometimes referred to as a *walk-through*, it is an informal session where the trusted agents and simulators talk the scenario through from beginning to end. This serves as a quality check to ensure the proper timing, integration, and flow of the scenario. It also allows for a final review of the technical aspects of the scenario before conduct.

Action Planning Session: The portion of a tabletop exercise that is the concluding activity. During this session, the facilitator and players identify lessons learned, action items, and next steps.

Scenario Background: This is the material provided to players (typically of a tabletop exercise) to provide a backdrop for the session. It usually contains information on the date, time, geographic location(s), weather, and the status of world, industry, and organizational affairs. Occasionally, scenario background information is provided to players during a drill or exercise as well.

Moves: This term refers to the narrative scenario "snapshots" provided as part of a single-scenario tabletop exercise, with several stages or updates over some period of time.

Conduct: This phase begins when personnel arrive for training and ends at the conclusion of drill or exercise play. Actual conduct activities differ dramatically between a tabletop exercise and a drill or exercise. This is explained further in Chapter 6.

Definitions of Positions/ Groups/Conduct Locations

Controller: Controllers are involved in drills or exercises. They monitor the flow of the drill or exercise. The role of a controller is to make certain that the drill or exercise is conducted in accordance with the scenario and timeline and within the established scope and parameters.

Facilitator: Facilitators are involved in tabletop exercises. They help lead the group discussion so that all relevant issues and questions from the scenario are raised and explored. Facilitators also lead the action planning session where players identify action items.

Evaluator: Evaluators are involved in drills or exercises and possibly tabletop exercises. Evaluators are responsible for observing player actions and evaluating the effectiveness of those response actions based upon good business management practices, defined objectives and associated response requirements, and evaluation criteria.

Simulator: Simulators are involved in drills or exercises. They simulate any person, organization, agency, or service that would normally interact with the responders during an emergency or crisis situation but who are not actually participating in the drill or exercise. Simulators act out these "roles" in order to provide the scenario information that drives the flow of the exercise and presents the postulated problems.

Simulation Cell: A simulation cell, sometimes referred to as a control cell, is used during a drill or exercise. It is an area or room set aside for simulators so that they may play out their roles in an integrated and coordinated fashion. The room usually consists of a bank of phones, a fax machine, and possibly a computer so that simulators can remain in constant communication with the players as they would during a real incident. Depending on the scope of the drill or exercise, there may be more than one simulation cell.

Simulation Cell Lead: The lead is the person responsible for managing the operations and information flow in the simulation cell to ensure play remains within the scope of the drill or exercise. To do so, the simulation cell lead must be in constant communication with the simulators as well as with the controllers so the scenario pace and content may be adjusted accordingly.

Player: A player is any individual or member of the response organization who is playing a defined response role during the drill or exercise activity.

Observer: Observers can be involved in both drills and exercises as well as tabletop exercises. Observers are invited guests who have no official role in the conduct of the exercise activity, though they are sometimes asked to submit their observations and insights from the activity. They are typically invited in order to encourage shared learning opportunities.

Drill/Exercise Coordinator: The drill/exercise coordinator is responsible for managing the entire conduct operation. He/she is assigned early in the design phase and is involved in every aspect of designing, developing, conducting, and evaluating a drill or exercise. The coordinator works closely with the designated lead controller to problem-solve and support effective conduct.

These are some of the more critical terms utilized throughout this book and in the context of exercising. A more complete list of terms and definitions is contained in the Glossary of this book. Again, it is important to accept these definitions within your organization and lexicon to the extent possible. The definitions support the foundation upon which this book's methodology is based—a methodology that encourages exercise planners to go beyond what has always been done and move toward the highest quality, most cost-effective process of validating your capabilities.

One final element of the foundation to be conveyed is the four basic phases involved in exercise activities. The four phases are: design, development, conduct, and evaluation. Each of these phases involves several distinct activities and tasks with associated timeframes for completion. Each phase has an entire chapter devoted to these activities.

Now that a common foundation has been laid for the book, you can turn your attention to assessing where your organization is in terms of readiness and capability. From there, you can begin developing a multi-year exercise program that is objective-driven and need based.

CHAPTER 2

Emergency & Crisis Management: An Assessment

The goal of crisis simulations is to make the unimaginable imaginable, and sensitize participants to the stress of crisis decision-making, and to identify and remedy individual, group and organization vulnerabilities under stress.

Jerrold Post, M.D.
Professor of Psychiatry,
Political Psychology and International Affairs
George Washington University

This chapter provides a discussion of both crisis and emergency management basics and a means of assessing your crisis management capability.

The objectives of this chapter are to help you

- understand the basic characteristics associated with crisis and emergency management and therefore the terminology used throughout the book,
- draw a mental picture of an "overall" response structure and how your organization "fits" into that response,
- evaluate the status of your capability and identify areas for validation and improvement through exercising,
- prioritize improvements to support exercise program planning.

Introduction

This chapter consists of two major sections. The first section is intended to provide players with a common understanding and background of the distinctions between a crisis and an emergency. The second section focuses on assessing your organization's current capability. An understanding of crisis and emergency management, as well as capability assessments, will support your exercise program planning and help you create more effective and realistic exercise activities.

Purpose

This chapter builds a base of knowledge for all readers. Understanding the characteristics of unanticipated events, the associated response requirements, and the overall response structure within which you will need to operate helps you to design realistic exercise activities. The capability assessment serves as the foundation for planning, training, and exercising. The assessment helps determine areas for validation and improvement in your capabilities and establishes priorities in your exercising program. The purpose of this chapter is

- to understand the distinctions between crisis and emergency management,
- to recognize vulnerabilities and their relationship to response requirements,
- to identify capabilities targeted for validation or improvement.

Defining Emergency and Crisis Management: Building a Common Understanding

The basic exercising terms defined in Chapter 1, *Exercising Basics*, are among those most frequently used in this arena; however, the list is not definitive (additional terms are defined in the glossary). As we noted previously, several of the terms and ideas utilized in this book indicate preferred definitions. There is no consistently used or common lexicon in crisis and emergency management among all industries, agencies, and organizations.

The following identifies the specific characteristics associated with each of the primary crisis and emergency management terms defined in Chapter 1. These are unique characteristics that should focus attention on associated response requirements.

Incident: Characteristics associated with incidents include:

- Incidents can occur fairly regularly and often do not escalate to the point of an emergency or crisis.
- Incidents should be mitigated as quickly as possible.
- Incidents can escalate to a worsening situation if not monitored and contained.
- Incidents can serve to identify prevention efforts which could decrease the likelihood of the incident occurring again.

Emergency: Characteristics associated with emergencies include:

- Immediate effects are often local in nature (e.g., on-site fire, pipe burst with oil spilling into containment dikes, or computer malfunction).
- Effects of the emergency have the potential to impact the site and/or surrounding community.
- Impacts typically are short-term (e.g., minor loss in production time, minimal air pollution due to fire smoke).
- On-site and/or local area resources are generally adequate to respond to and manage the situation.
- May or may not have potential to escalate into a crisis.
- Timeframe for action is extremely limited.
- Initial information regarding the situation is limited and at times conflicting.

Crisis: Characteristics associated with crises include:

- Potential to impact the organization with long-term consequences, including possible short- and long-term ramifications for strategic and business plans.
- May require significant corporate resources (financial support, staffing, management and technical expertise, etc.) to respond effectively.
- Off-site and contract resources are usually required.
- Timeframe for action is extremely limited; however, the impacts of the situation may continue long after the situation is brought under control.
- Limited and often conflicting information regarding the situation.

Disaster: Typically, disasters are thought of in terms of natural disasters. As a result, the consequences of these events are thought of in terms of impacting a specific geographic area. Recent disaster examples include: Hurricane Andrew (1993); Mississippi River flooding (1994); and Kobe, Japan, earthquake (1995).

Characteristics associated with disasters include

- broad geographic impact;
- increased demand for limited available resources in the area such as fire response, medical support, etc.;
- communications ability typically impaired;
- restrictions and/or limitations on mobilization and movement in the area;
- potential loss of basic utilities and damage to infrastructure;
- broad-based impact on area employees that could cause extended staffing shortages.

Emergency Management: The focus of emergency management and incident command tends to be tactical, i.e., to "put out the fire" and manage all the activities associated with that task. For example, an emergency manager or incident commander should be considering and managing the following tasks:

- notifying response personnel an incident has occurred requiring immediate protection of life, property, and the environment;
- managing emergency resources and response activities at the site (i.e., fire brigade, evacuation teams, etc.) and
- ensuring that necessary equipment is either in place or en route to the site.

Crisis Management: Crisis management encompasses tasks associated with overseeing the management of emergency response; however, the focus is on the consequences and strategic aspects of the incident. A crisis team usually has oversight authority with regard to the emergency management efforts.

The strategic and long-term consequences of the incident should be the primary focus of a crisis management team. In the fire example used previously, the crisis team should be considering such issues as

- the long-term impact on company image,
- the need and parameters of humanitarian or claims efforts for the surrounding community,
- the impact of the crisis on a pending contract to build a similar facility in the town 10 miles away,
- other long-term business consequences that may arise from the situation.

The common understanding of these terms enables the organization to determine how many levels of response may be required. The next section of the chapter discusses the various organizational structures and different roles and responsibilities.

The Overall Response Structure

Effective management of incidents, emergencies, and crises depends upon a number of factors. We have identified four general characteristics for an effective response. They are (1) a clear understanding of how you define an emergency and crisis, (2) pre-identified response roles and responsibilities, (3) well-trained and well-exercised personnel, and (4) flexibility.

1. Clear understanding of emergency and crisis: It is essential to define *in advance* what constitutes a crisis along with the level of response and extent of resources needed for an effective response. Correctly identifying the scale of an incident is an integral first component to triggering the appropriate response. It also helps ensure incidents are appropriately classified and reported, and that support is available to meet worst-case scenarios.

Distinguishing between an incident, emergency, and a crisis is also important from a procedural point of view. Organizations should focus on the development of programs that specifically address the resolution of events in each category. Decision-makers can use the distinctions to guide which aspects and levels of the program are appropriate to implement for a given event and to anticipate needs.

2. Pre-identified response roles and responsibilities: Internal response groups, from the field level through senior management and functional support teams such as human resources, should be identified in advance of an incident. Ensure your structure and staffing takes into consideration a series of scenarios including those ranging from an incident to a crisis. This should help ensure that adequate personnel and resources are available during a worst-case situation or crisis. Response groups external to your own (such as government agencies/officials if you are a business, or media and outside contractors for any organization) should also be identified prior to an incident.

The roles and responsibilities of each internal and external group, at all levels of response, should be well thought out in advance of a response situation. Through training and exercising, further refinements of these roles and responsibilities will be made. Unless training and exercising activities that address a range of situations are conducted over time as organizations and personnel change, the basic planning and response assumptions will likely prove to be invalid during an actual event.

3. Well-trained and exercised response personnel: Personnel with a designated response role should also participate in training and should know, and be prepared to fulfill, their response roles and responsibilities. Responding to an incident usually involves interaction with any number of internal groups and perhaps one or more external groups. Having previously trained, exercised, or otherwise interacted with these groups will enhance the response effort during an actual event. Additionally, the various personnel and groups will be able to begin functioning more quickly in a coordinated, team-like effort to respond to and manage the situation, if they are well trained and practiced.

4. Flexibility: An organization should have the ability to expand and contract the response effort as necessary. An emergency or crisis manager must be able to expand response through requesting additional support teams, equipment, and other resources to best meet the potential needs of the situation. Initially, the worst probable (or planning) case should be assumed and the resources necessary to manage and mitigate the effects of that "planning" case should be mobilized.

Expansion and contraction does not end at the site level. Organizations should ensure the event is being evaluated and monitored at all levels. Through monitoring, it is possible for the organization to remain ahead of, and anticipate, response needs. For example, if an incident has the potential to impact citizens off-site, the emergency team should notify local officials of that potential and help to ensure protective actions can be taken in ample time.

Contraction of the response effort is also important. Once it becomes clear certain resources are no longer needed, those resources should be taken out of the active response operation. The remaining response organization and resources should be able to continue functioning effectively as the total effort contracts.

Internal Organizational Structure

The organizational structure is an essential component of crisis and emergency management. It is the foundation of your capability. Generally, organizations have three or more levels of response. Although most levels include multiple subdivisions, the three most common levels are field level response, emergency management support, and crisis management.

A defined organizational structure helps to ensure that responders understand their roles and areas of responsibility. Required or expected interactions and coordination among groups is preestablished through this structure, and as a result, simplified. In many industries, regulatory agencies are providing guidance or, in fact, dictating certain aspects of your response organization. These requirements, such as use of an incident

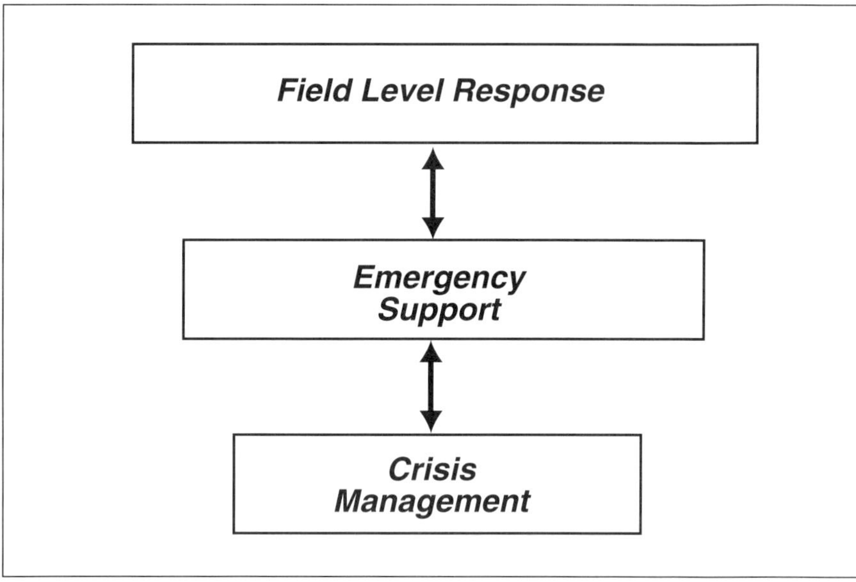

Figure 2-1
Internal Organization Structure

command system and the unified command concept, are consistent with what we are describing as three levels. The levels of an organizational structure, including an example of personnel within that level and corresponding areas of responsibility, are depicted in Figure 2-1 and further described here.

Field Level Response

The field level typically represents the levels of an organization that are closest to producing, delivering, or selling your product or service. For example, in a manufacturing/industrial organization the field level would include facilities or plant operations. In the financial sector, the field level would likely involve a loan servicing or data center. Field level personnel are responsible for prevention and mitigation of incidents: planning and taking the tactical response actions. As the first line of defense or response to an incident, personnel at this level are typically responsible for performing incident or emergency response functions within the immediate geographic area and undertake response operations such as "putting out the fire." This level may encompass both field teams and emergency management teams. It also includes what is typically referred to as incident command. The emergency support teams may or may not be considered part of the field level response depending upon the culture of the organization involved. Emergency support is described in more detail here.

Field level personnel can include facility or technical response personnel (e.g., incident commander), special teams (e.g., fire brigade), local area resources (e.g., police, emergency medical technicians), and area response personnel (e.g., mutual aid and cooperatives).

Emergency Support

Emergency support typically refers to the middle level of many organization's response structures. The team(s) at this level are referred to by different names, depending on the industry or your organization. For example, many organizations call this team the Incident Support Team (IST), Emergency Management Team (EMT), or Management Emergency Response Team (MERT). Regardless of the name, the primary function of this team is to provide support to field response personnel. This support may be provided from the field level (if the site is large enough) or from the region or headquarters level. It may also be provided by some combination of teams at each level. This team also serves in an informational role, often by providing incident reports, incident evaluations, and recommending actions to responders and/or senior management. By monitoring both the incident and the response, this team can anticipate and recommend actions and resource needs. Actions may include dispatching special teams to the site and/or suggesting that senior management become involved to support overall response.

At the emergency support level, response roles tend to be functional or more specialized. Emergency support personnel can include specialized teams such as humanitarian assistance teams, spill response strike teams, and business continuity teams. Functional representatives such as procurement experts, claims personnel, and communications experts are also included on emergency support teams. Operations and technical personnel are also frequently a part of the emergency support level.

Emergency support roles can include

- anticipating, recommending, and providing needed resources and support;
- coordinating the provision of resources from different units within the organization;
- monitoring of the incident and its potential to become a crisis;
- providing essential information to senior management regarding the incident and the response;
- providing recommendations and/or carrying out business continuity efforts;
- managing any other support activities such as media requests, meetings with authorities, and supporting recovery planning efforts.

Depending on your organization, the emergency support level of response may occur near the site or in the affected facility, or it may be accomplished largely at your organization's headquarters or other offices remote from the scene.

Crisis Management

The crisis management (CM) level tends to involve the most senior level of the organization. This level is typically comprised of a multi-disciplinary team made up of both line and functional senior managers. For example, the CM level can include senior managers from essential departments or functional areas, such as environment, health and safety, law, etc., and senior managers of operating groups or functions (e.g., manufacturing, engineering, loan processing, information technology, etc.) The crisis management level becomes involved in incidents that have the potential to seriously affect the overall way an organization conducts its business or the organization itself—for example, an organization's image, market share, public/regulatory standing, and financial stability.

The primary role of the crisis management level is to evaluate and mitigate the strategic and potential long-term business consequences associated with a crisis.

Crisis management roles can include

- evaluating the overall response to, and possible consequences of, an incident;
- ensuring adequate support is available to the overall response effort;
- taking actions appropriate to minimize both long- and short-term impacts of an incident;
- evaluating long-term consequences of an incident on the organization;
- developing strategies to prevent or minimize the consequences of the incident.

External Organizations

The "external world," including government agencies and other organizations outside of yours, can also play a role in response. When assessing external parties, your organization should take into account the available resources these groups can offer, in addition to those external groups you may need to coordinate and cooperate with during a response. Figure 2-2 depicts the various groups who might be considered as external organizations requiring interaction with your response organization.

External groups that can typically provide response resources include local fire and police departments; emergency clean-up, data recovery (hot-sites), and mitigation contractors; volunteer personnel; mutual aid

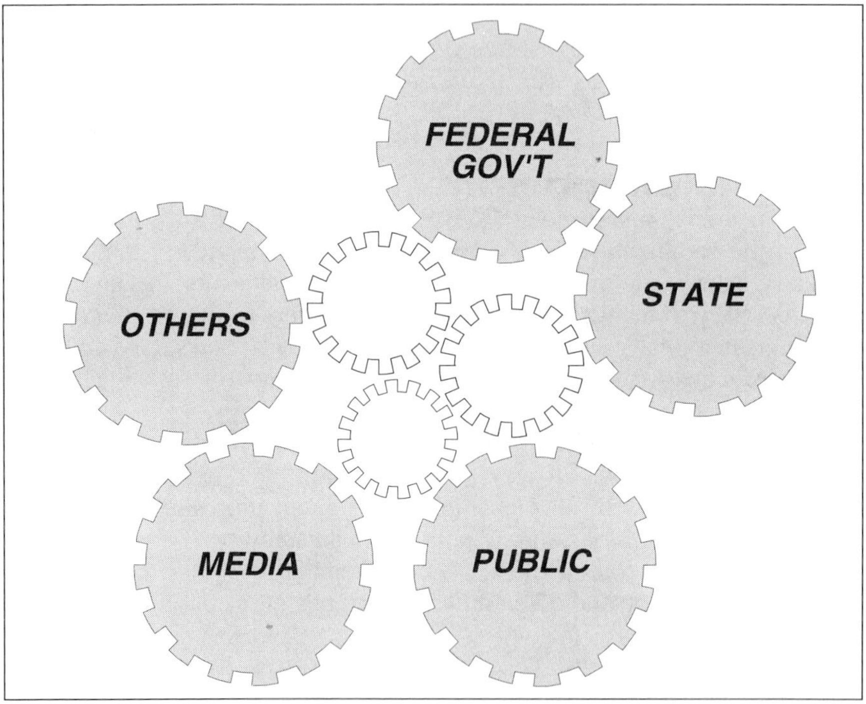

Figure 2-2
External Organization Interaction

groups, cooperatives, and area response teams; local, state, and federal government response personnel (e.g., EPA, USCG); and contracted equipment and personnel. A full assessment of the resources these groups actually offer may suggest that they be built into your own response planning.

In addition, interaction and coordination may need to occur with government officials, regulatory personnel, media personnel, the community, employees and employees' families, and other stakeholders such as environmental groups, city officials, and industry associations.

Where is your focus?

It is important to determine your focus within the big picture of crisis and emergency management. No individual, department, or team can, or should, try to be self-sufficient during a crisis or emergency. To ensure effective response, each component of an organization must understand its interdependency with other groups, both internally and externally. Refocus your attention to encompass the overall response organization and determine the groups with which you need to interact and coordinate. Initially, identify those groups most dependent upon your support and vice-versa.

Continually expand your view until it includes any group you would potentially coordinate and/or communicate with during a crisis. This big picture view will allow your organization or group to clearly see how it "fits" into the total response capability and what its focus is.

To better identify your focus and your circle of coordination, consider the following:

Internal

- Identify your group's primary roles and responsibilities as emergency- or crisis-related,
- Evaluate the organizational structure above your group,
- Evaluate the structure below your group,
- Evaluate what your expectations are from these different groups,
- Define the roles and responsibilities of your group vis-à-vis the other groups you have identified.

External

- Evaluate and identify what regulatory agencies, government officials, etc., you believe would become involved in an incident;
- Determine the potential for high media exposure of an incident (and what media type it would be, i.e., radio, television, newspaper);
- Evaluate mutual aid contractors, industry associations, and other external resources that might play a role in response;
- Assess the actual interaction and coordination among groups based upon previous incidents or history.

Assessing Your Crisis Management Capability

In building a foundation for an effective exercise program, you begin by ensuring a clear understanding of how your organization fits into an overall response picture. However, to complete our foundation, this chapter focuses on assessing your organization's current capability, i.e., how well is your organization prepared to manage an emergency or crisis today?

Overview of Capability Assessment Steps

A capability assessment should enable you to judge how prepared you are to manage and mitigate effectively the impacts of an incident. The capability to respond to an incident quickly and effectively should lead to

- a lessening of the incident's impacts on the surrounding population, environment, employees, etc.;
- positive (or maintained) relations with the impacted communities;
- positive perception of the company's response efforts.

To assess capability you should assess five primary areas. Through an evaluation of these areas you should gain a strong indication (or snapshot view) of your organization's capability. This capability assessment is not designed to replace other, more technical assessments, but rather to build upon them. The five major steps in the capability assessment are

1. identification of vulnerabilities,
2. assessment of fundamental capabilities,
3. assessment of other responding parties,
4. assessment of resources,
5. assessment of your maintenance program.

Your evaluation will provide a "snapshot" view of the existing capability in your organization. Because organizations are continually evolving, a capability assessment should be considered every two to three years. The process described here provides for a macro-view and summary of each area that is typically based on more detailed/extensive reviews of these areas. The graphic shown in Figure 2-3 shows the steps in assessing your capabilities.

Step 1: Identification of vulnerabilities: It is essential that your organization identify its vulnerabilities. Determining vulnerabilities allows your planning to reflect required response to those incidents that are either likely to occur or those incidents less likely but which would cause a major disruption to normal operations. Vulnerabilities represent the impacts and consequences of hazards. Hazards are any source of danger, which may have impacts and therefore consequences for the organi-

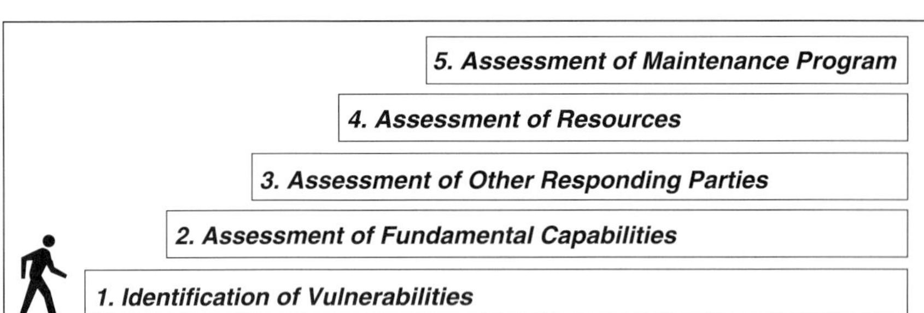

Figure 2-3
Five Major Steps of Capability Assessment

zation. In order to identify vulnerabilities, you should look at both hazards and the potential associated impacts of those hazards on your organization. Attempt to categorize and summarize hazards to assemble a complete view of the vulnerabilities facing your organization. You may be required to gather information from a number of locations or operations to appropriately identify the range of vulnerabilities your level of the organization may face.

The following outlines three categories of hazards and provides short definitions and examples. Figure 2-4 provides a more complete list for your consideration. As you review these hazards, also consider the impacts.

1. BUSINESS OPERATIONS

- chronic environmental contaminations
- computer equipment or software failure
- disruption of raw material procurement
- equipment failures
- explosion
- fire
- industrial accidents
- toxic material release or spill
- transportation accidents

2. NATURAL DISASTERS

- earthquakes
- epidemics
- floods
- hurricanes
- severe winter weather
- tornadoes/severe thunderstorms

3. EXTERNAL DISASTERS

- incidents in geographical area of facility
- utility/telecommunications disruptions
- "similar industry" incidents
- "other industry" incidents

4. OTHER

- Product-related
  - design or manufacturing defects
  - contamination
  - tampering
  - theft
  - trademark/copyright infringement
- Employees
  - accidental death
  - personal injury
  - chronic safety problems
  - embezzlement
  - sabotage
  - labor strife
  - sexual harassment
  - violence in the workplace
- Management
  - downsizing/reorganization
  - executive misconduct/fraud
  - executive succession
  - new business/foreign investment
- Legal/Regulatory
  - citations by occupational or environmental agencies
  - lack of legal framework/precedence/enforcement
  - local/regional/country actions
  - nationalization/expropriation of assets
  - remediation, fines, penalties
  - restrictive legislation/judicial decisions
- Financial
  - competitor actions
  - divestitures/joint ventures
  - hostile takeover attempts
  - market share loss
  - stock market/exchange rate/price collapse
- Societal
  - civil unrest
  - consumer protest/boycott
  - kidnapping
  - political instability
  - terrorism/hostage-taking

Figure 2-4
Common Hazards Faced by Many Organizations

Operational Hazards: Processes and other operational functions related to the work environment that have the potential to cause an emergency and/or crisis for the organization.

Examples include

- computer viruses or system failures,
- fires/explosions,
- toxic material releases,
- product contamination or tampering,
- environmental contaminations,
- violence in the workplace,
- workplace deaths/injuries,
- employee dishonesty such as embezzlement or theft.

Non-Operational Hazards: Incidents not directly related to processes and products but with the potential to cause an emergency or crisis.

Examples include

- executive succession or kidnapping,
- regulatory citations,
- labor relations,
- financial strain,
- hostile take-over attempts,
- lawsuits for discrimination, sexual harassment, etc.

Natural Hazards: Naturally occurring events that cannot be prevented—acts of God. Natural hazards include considerations of weather patterns and other phenomena that are likely to occur in your organization's location or operating areas.

Examples include

- hurricanes and tsunamis,
- floods/droughts,
- earthquakes,
- severe winter storms/conditions.

Beyond the categories of hazards identified here, history shows that actual incidents often involve an initial event triggered by external forces. Moreover, an incident can be compounded by other or subsequent events. For this reason, you are encouraged to use a technique called "What if" analysis. Similar to brainstorming, in a "What if" analysis you and others assist in assessing your organization's capability by exploring the full range of possibilities by asking what would happen if. This process should allow for any idea to be considered, regardless of how remote its possibility of occurrence may seem. The results usually uncover some events and impacts not previously considered or recognized.

Step 2: Assessment of fundamental capabilities (personnel, structure, and systems): Fundamental capabilities are primarily internal resources and structures. The assessment should focus on the type of organizational structure in place, the concept of operations, staffing, and coordination mechanisms. While you are reading through the following pages, ask yourself "to what extent does my organization... ." Answering the extent to which you have trained personnel, the extent to which you have a documented plan, etc., should provide a fairly clear indication of the status of existing capability. You will be making, in some cases, judgment calls or educated guesses as to the extent of the capabilities. It is important to be honest in stating the *real* capabilities. Be careful not to overstate what your organization is actually capable of or how well it will perform in a real event. It is often useful to rely on actual past events or exercise activities to assess these capabilities.

Organizational Structure: The organizational structure refers to those individuals, teams, departments, functional specialists, etc., that comprise an organization's internal crisis management structure. A defined structure that identifies crisis and emergency management functions and specialized teams should be established and understood throughout your organization. Consider the duration of the expected response functions and the number of personnel and designated alternates available to support needed operations. Assess the extent of existing skills and training completed by both individuals and teams.

Concept of Operations: The primary purpose of a concept of operations is to outline planned response actions and intended interaction among all responders. This concept serves as the game plan during actual response, i.e., who is going to do what, and how the response efforts will flow and integrate. The concept of operations should help ensure all responders clearly understand their role as well as their relationship to the rest of the organization and external groups.

The concept of operations should address

- emergency response/crisis management policy, guiding principles, and standards;
- reporting procedures and systems;
- notification requirements and expectations for internal and external groups;
- guidelines on initial information to be provided;
- any identified functions for individuals and teams with a response role;
- the defined command structure and planned interactions;

- built-in flexibility to expand and contract response operations as needed;
- communication flow (process) requirements and systems;
- integration with internal and/or external groups;
- use of facilities and equipment;
- staffing.

By addressing each of the above categories in the concept of operations, the organization is ensuring a more effective response. The more comfortable responders are with this "game plan," the better prepared they are to response in the desired manner.

It is important to understand that an effective capability is dependent upon people. Thus, having a well-qualified, well-trained staff is essential. Having an appropriate depth of personnel assigned to each designated response position also is essential. All personnel assigned to response roles should have the appropriate background, level of authority, and training to enable them to carry out their roles and responsibilities. It is the combination of experience and regular training that assures a constant state of readiness.

Communication and Coordination: Communication and coordination are essential to ensuring an effective response. Associated requirements should be established in advance, planned for, and addressed in training. Identified communications and coordination requirements for internal as well as external groups (government, media, industry associations) should be considered and factored into planning for the overall capability. Appropriate equipment and materials, with adequate back-up capability, should be provided to support communications.

Step 3: Assessment of other responding parties: Response to an incident does not occur in a vacuum; as a result the requirements of other responding parties should be considered in your capability assessment. For example, government agencies, such as the United States Coast Guard (USCG) or the Environmental Protection Agency (EPA), have regulatory requirements to perform some response functions. Other groups that have agreed to support your organization's response through prearranged mutual aid and/or contractual agreement should also be considered. Additionally, others agree to support response efforts through recognized volunteer organizations, such as the Red Cross.

The requirements often result in the organization and the external group responding concurrently to an incident. Your organization should attempt to maximize efficiency, without decreasing effectiveness, through ensuring that the roles of other parties are fully considered when developing your response strategies.

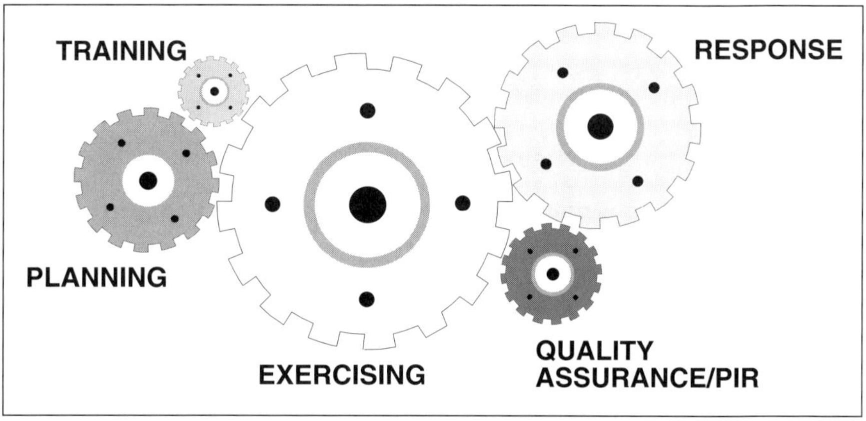

Figure 2-5
Crisis Management Program

Step 4: Assessment of resources: In order to carry out your organization's planned response, resources and tools are needed to support the effort. Every resource that can help or support your organization in response should be identified and included in planning and, as appropriate, training.

Resource assessments should review

- known resources available at each site;
- equipment and other resource needs, and lists of where the resources can be obtained;
- emergency systems for procurement, delivery, and replacement of supplies and material;
- other available personnel, equipment, or materials provided through mutual aid agreements and contractor agreements established prior to an incident.

Step 5. Assessment of the maintenance program: Your organization's capability is maintained through program activities such as planning, training, self-assessment and quality assurance programs, exercising, and the conduct of post-incident reviews. These efforts support capability by providing a program foundation and continuous feedback and validation. Each component of your program should be integrated as shown in Figure 2-5. Similar to gearing up your exercise program, each of the other activities mentioned needs to be geared up as well. Proper functioning of the overall crisis and emergency management system requires that all the gears work together and turn at compatible speeds. The gear for planning may turn at a rapid rate as new plans and response roles are delegated. The

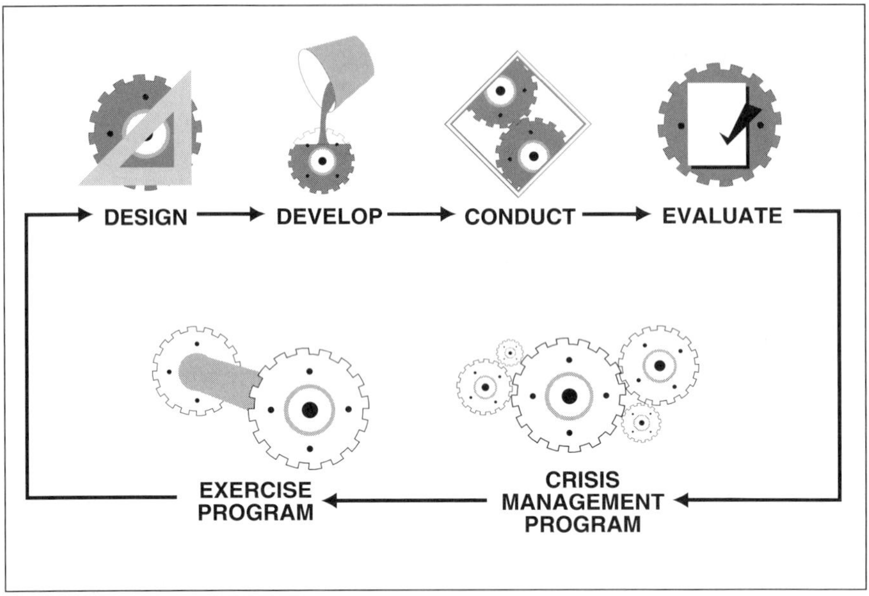

Figure 2-6
The Exercise Cycle

training gear needs to move at a compatible rate to ensure it keeps pace with modifications to response roles and so on.

Conclusion

This program and assessment chapter should provide a base of knowledge and a common foundation for the remainder of the exercising book. Armed with a common understanding of crisis and emergency management terms, an identification of your response structure, and most importantly, an understanding of how the group for which you are responsible fits within the overall response structure, you are ready to program your exercise activities more effectively. You should now be prepared to integrate the results into your exercise program.

The overall exercise cycle, encompassing your crisis management program, exercise program, and the individual phases of each exercise activity (design, development, conduct, and evaluation), is depicted in Figure 2-6. This cycle is the basis of this exercising book.

HOW TO

Assess Your Capabilities

- ☐ 1 *Identify Vulnerabilities*
  - *List risks*
  - *List vulnerabilities*
  - *Prioritize based on likelihood and/or severity*
  - *Summarize vulnerabilities*
- ☐ 2 *Assess Capabilities*
  - *Identify the emergency and crisis management organization*
  - *Outline the concept of operations*
- ☐ 3 *Assess Other Responders*
  - *Identify external response capabilities*
- ☐ 4 *Assess Resources*
  - *Inventory internal staffing, equipment, facilities and other resources*
  - *Inventory potentially needed external resources*
- ☐ 5 *Assess Maintenance Program*
  - *Assess training program/records*
- ☐ 6 *Review Exercise Program and Past Evaluation*
- ☐ 7 *Review Any Incidents*
- ☐ 8 *Compare Results to Needs*
  - *Existing capabilities by component*
  - *Existing capabilities of overall organization*

CHAPTER 3

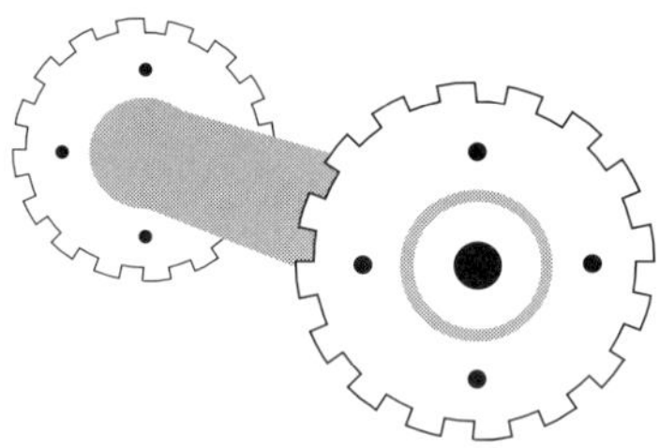

EFFECTIVE EXERCISE PROGRAM

Undertaking a large-scale exercise prematurely, or without adequate context, can be a frustrating and expensive disappointment. A full-scale simulation should represent one aspect of continuing, incremental crisis training–not an attempt to shortcut or substitute for the overall process.

Mary Woodell
Senior Principal
CMRC & Co.

This chapter describes the elements of an effective multi-year exercise program and leads you into a process to accomplish exercise program planning.

The objective of this chapter is to help you understand how to plan a multi-year exercise program.

Introduction To Exercise Programming

Exercise programming builds upon the concepts of crisis and emergency management and the capability assessment to help determine your goals and objectives for exercising. In this chapter we discuss different types of exercise activities and describe the process of developing a multi-year exercise program.

The Multi-Year Exercise Program

Managing a business operation requires strategic planning and forward-looking programs; managing your exercise program is very similar. An effective program is based upon goals and objectives and on an understanding of current capabilities versus the desired level of capability. The process is systematic and lends a methodical approach to your exercising efforts.

Developing a multi-year exercise program uses a building-block approach to

- establish a framework for gradually improving capabilities (generally based on some type of capability assessment),
- promote integration of capabilities through a multi-year planning process,
- ensure proper timing in exercise planning to capitalize on past lessons learned and to allow time to make improvements,
- allow more effective budgeting and use of resources while maximizing learning value through proper exercise programming.

Characteristics of Exercise Activities

As previously mentioned in Chapter 1, a number of different terms are used throughout the industry to define the various exercise activities. These activities fall into one of three categories: *Tabletop Exercise*, *Drill*, and *Exercise*. In order to outline an exercise program, we need to take each exercise activity and analyze its characteristics. This analysis will enable us to understand when to use each of the various activities and why. The characteristics will also demonstrate the benefits and potential uses for each type of exercise activity in your multi-year exercise program. The characteristics associated with each exercise activity are further described here.

Tabletop Exercise: The tabletop exercise is sometimes referred to as a "desktop exercise." These can be low-cost, highly valuable activities for your organization. Characteristics include

- a scenario-driven discussion (no physical response);
- limited to a small team or group, usually in a conference room setting;
- used for establishing policies, providing input to planning, developing procedures, clarifying roles, and confirming knowledge;
- a nonthreatening learning environment;
- least "realistic";
- the uses of specifically trained personnel to facilitate discussion and evaluate performance;
- a four- to six-week process, from design through evaluation.

Drill: Drills are action oriented and more extensive in terms of cost than a tabletop exercise. The planning process is also somewhat longer. There are several levels of drills and the characteristics associated with each type are identified below. Drills enable an organization to validate key functions and demonstrate individual and team knowledge, skills, and capabilities. In general, drills involve

- scenario-driven response activities;
- limited number of groups playing;
- simulation of nonparticipating parties;
- limited mobilization of personnel and resources;
- limited scope;
- focus on specific areas of crisis/emergency management;
- hands-on experiential learning;
- limited or no validation of interaction and coordination among a group(s);
- trained personnel to serve as controllers, evaluators, and simulators;
- a six- to nine-month process, from design through evaluation.

Limited Drill: The limited drill is utilized to evaluate capability in a very specific area or function. This type of drill typically involves a single function or facet such as equipment tests or notification and callouts.

Full Drill: The full drill, in contrast, is utilized to evaluate capabilities for a range of functions. Although still limited in scope, this type of drill typically involves full team participation and looks at limited interactive response activities among response levels and/or groups (internal or external).

Exercise: Exercises are designed to mirror a real-life incident as realistically as possible. This type of activity enables the organization to validate a multitude of functions and teams as well as to evaluate the interaction and coordination among all the various players. Exercises involve

- scenario-driven response activities;
- a range of size and scope of participation;
- simulation of nonparticipating parties;
- a range of personnel and resource mobilization;
- participation of multiple teams/management levels in *integrated* response;
- hands-on experiential learning;
- comprehensive validation of interaction and coordination within a system;
- encompassing and increasingly complex scale and scope;
- extensive realism;
- trained personnel to serve as controllers, evaluators, and simulators;
- a 9- to 12-month process, from design through evaluation.

Elements of an Exercise Program

Building an exercise program requires that an infrastructure be in place to support exercise activities. The infrastructure identifies the methodology, systems, and approach used in support of the exercise process. The infrastructure should also include the procedures for maintaining and supporting these program efforts. First, it is essential that responsibility be clearly assigned to administer the exercise program. The person(s) assigned responsibility should ensure that

- overall program goals and objectives are defined;
- a budgeting process is in place and utilized;
- the annual, and preferably multi-year, exercise schedule is coordinated and integrated;
- adequate support is provided throughout the organization for the effective design, development, conduct, and evaluation of individual exercise activities;
- a resource management system is employed to maximize efficient use of resources organization-wide;

- a lessons learned tracking and sharing system is in place and there is an established process for using it.

Multi-Year Exercise Program Planning

Exercise planning in multi-year time frames ensures program as well as individual team goals and objectives are being identified, and that a means for reaching those goals is determined. Planning for a multi-year timeframe also enables the organization to identify areas where the goals of two or more groups can be met by combining an exercise activity. The fundamental purpose is to establish the desired level of capability and to plan a systematic process for reaching that goal. A complete timeline for each activity, from design through evaluation, should be planned. Sufficient time between activities should be allowed on the master schedule for improvements to be initiated and put into place.

Finally, the multi-year plan enables the organization to maximize the learning opportunities of each activity by planning involvement of other parties (i.e., observers, trusted agents, etc.) and by sharing the lessons learned with a wide audience. It can also spread expenditures more effectively over a longer period of time while continually building and validating response capabilities.

Developing a Two-Year Plan

Set exercise program objectives: The overall exercise program objectives should be determined based upon your capability assessment. These macro objectives should identify the capability desired, which can suggest the path by which the organization intends to reach that capability.

Consider recommended frequencies for exercise activities: Frequency of activities is needed to ensure teams and individuals maintain understanding and proficiency. The graphic in Figure 3-1 and the text following identify the recommended frequency of exercise activities.

- Depending upon the nature of the roles and responsibilities being exercised, the frequency of exercise activities should be adjusted. For example, notification operators should be exercised more frequently than senior management teams, based on the likelihood of their involvement in an actual incident.
- Team interaction should occur at least once every three to four months. The format of these interactions does not have to be simulation activities, i.e., quarterly discussion meetings are satisfactory. However, some type of interactive activity once per quarter encourages team building and ensures that those

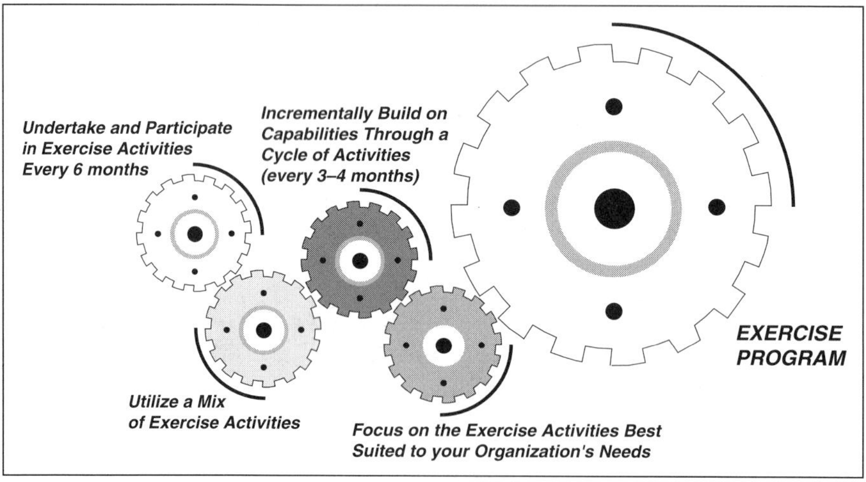

Figure 3-1
Exercise Activity Scheduling Guidelines

individuals serving on teams are comfortable with their roles and responsibilities as well as comfortable with one another.

- Simulation activities every six months, or approximately twice a year, helps ensure capabilities are maintained. More importantly, regularly scheduled simulations enable your capability to increase systematically through identifying areas for improvement and a subsequent validation of those improvements.
- Scheduled exercise activities should allow time between activities to update/revise plans and procedures and make the suggested modifications to processes.
- Exercise activities should ensure that recently adopted procedures and new personnel are brought "up to speed" through scheduled activities.
- A mix of exercise activities, from tabletop exercises to full-scale exercises, should be used by the organization to build upon capabilities in increments. The organization should use each type of exercise activity at different points in the capability development and validation process.
- Some exercise activity types are better suited to certain team compositions, functional assignments, and organizational levels. Select the best mix of activities for each team and the organization overall. A sample exercise program schedule is shown in Figures 3-2a and 3-2b.

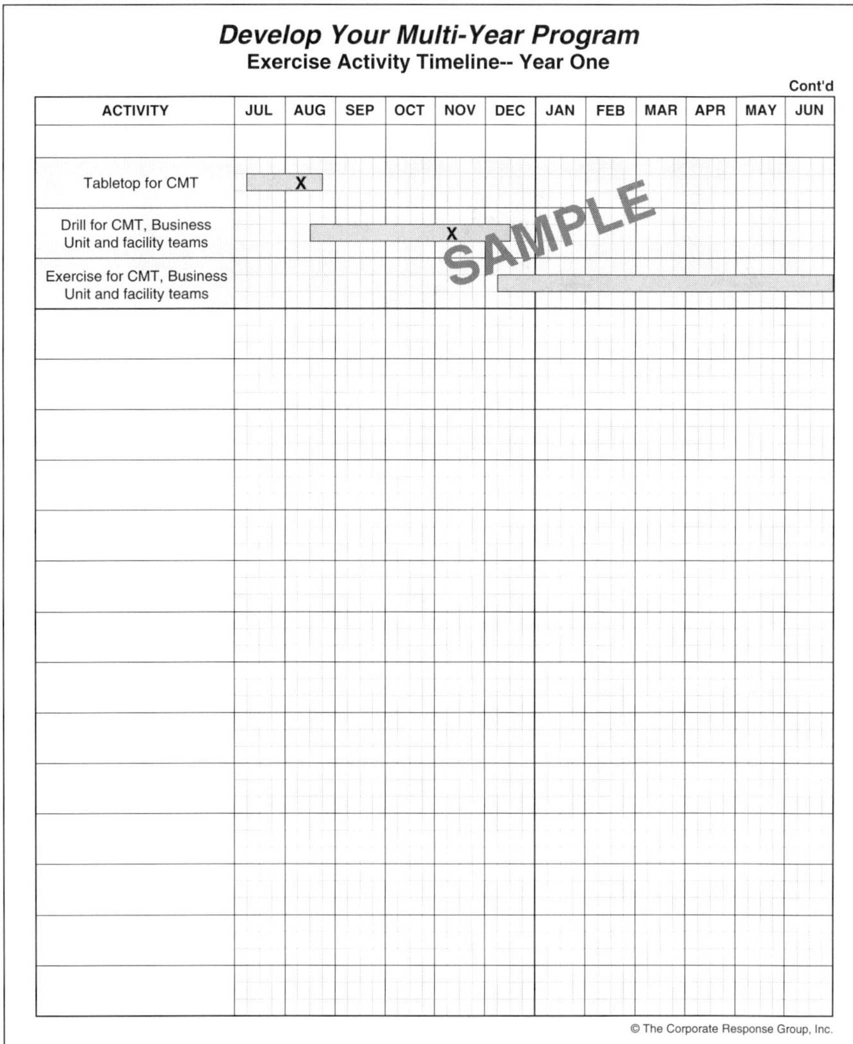

Figure 3-2a
Develop Your Multi-Year Program, Exercise Activity Timeline—Year One

Critical Factors: A number of different factors should influence your decision to schedule exercise activities for individual groups. Some of the considerations are

- What development or validation needs do the organization, teams, and individuals have?

Develop Your Multi-Year Program

Exercise Activity Timeline-- Year Two

JUL	AUG	SEP	OCT	NOV	DEC	JAN	FEB	MAR	APR	MAY	JUN	ACTIVITY
												Tabletop for CMT
												Drill for CMT, Business Unit and facility teams
X												Exercise for CMT, Business Unit and facility teams

Figure 3-2b
Develop Your Multi-Year Program, Exercise Activity Timeline—Year Two

- Where do capabilities need to be improved or revalidated? To answer this question you should consider your capability assessment(s) and the following factors:
    - level of understanding of roles and responsibilities among personnel;

- level of training completed by personnel and their last training opportunity;
- number of new personnel;
- last team activity of any kind;
- assumed and validated level of capability;
- best approach to build capabilities gradually, based on performance requirements;
- last actual performance of capability (exercise or real event).
- What regulations exist that require exercise activities of certain teams, operations, or levels of the organization?

Timeline: Several considerations are essential to developing a two- or multi-year timeline of exercise activities that will enable you to meet your exercise program objectives. While developing your exercise program plan, you should

- consider simulation activities every six months;
- utilize a mix of exercise activities (ensure the amount of time noted for completion of each activity is accounted for in the schedule);
- build on capabilities incrementally, through a cycle of activities;
- focus on group activities best suited to the makeup and needs of the particular group;
- ensure any timing identified in regulatory requirements for exercising is considered.

Although exercising is the focus of this book, your organization and exercise program planning should be integrated with other scheduled crisis and emergency management activities. For example, some of the team building necessary to maintain an effective and cohesive response team can be accomplished through frequent (monthly to quarterly) meetings. However, it is important to ensure a crisis or emergency management focus at these meetings and, to the extent possible, a similar meeting and style process to that of an actual response environment. Linking your exercise program to the rest of your crisis and emergency management program is discussed in more detail in Chapter 10.

Integration Opportunities

Integrating exercise activities of teams throughout the organization helps increase exercise activity effectiveness without increasing the cost. Once the objectives of exercising have been determined, the exercise program manager(s) should look for opportunities to integrate the activities.

This is accomplished by linking teams and/or organizations with similar goals into one exercise activity. In order to achieve these benefits, the two- or multi-year plan must identify the objectives of all groups as well as the relationship of these groups. Both internal and external parties should be considered and included in your process. Where regulatory requirements outline exercise activities, external involvement of players or evaluators should also be factored into your integration planning.

As an integration example, your two-year plan may identify a field-level team objective to be the validation of its notification capability. The timeline also indicates that the notification operators have an objective to validate their new information-gathering sheets. Both objectives can be accomplished with one exercise activity. It is important, however, to keep an overall organizational perspective to ensure that every activity does not become joined into only a few longer-scale activities. A balance must be struck between individual, focused activities and integration of your program's components in longer exercise activities.

We have discussed the various considerations of exercise activity and program planning. At this point you should be comfortable with the different types of exercise activities, the time involved with each activity, and the overall program planning process. Based on your exercise program plan, you are now ready to create actual exercise activities following the four exercise activity phases.

HOW TO

Create Effective Exercise Programs

- ☐ List capabilities, functions and/or teams to be validated or revalidated following improvements.
- ☐ Identify exercise program objectives.
- ☐ By team, subgroup, or individual position, identify (1) objectives and (2) exercise activities best-suited to accomplish these over two or more years.

  Note: Keep in mind the purpose and value of each exercise activity type to identify the rationale for your selections.
- ☐ Draft a timeline of exercise activities by team (reflecting time requirement for each).
- ☐ Look for scheduling and integration opportunities based on timing, objectives, and exercise activity type.
- ☐ Adjust exercise program timeline to reflect integration.

CHAPTER 4

Maximizing Value Through Design

One of the principal keys to a successful simulation is the amount of care going into its design. A scenario that is realistic down to the finest points will lay the groundwork for success.

Mayer Nudell
Author
Handbook for Effective Emergency and Crisis Management

This chapter lays the groundwork for designing an effective exercise activity.

The objectives are to help you

- increase your knowledge about elements of designing any exercise activity;
- prepare for your activity of designing a tabletop exercise, drill, or exercise.

Introduction

Based on your exercise program, your organization has selected exercise activities by team (or number of teams and/or groups) best suited for their capabilities and needs. For each activity you should also decide the extent to which you will address validation versus training and how you plan to maximize learnings and continuous improvement opportunities.

Now you are ready to go through the process of designing an exercise activity.

Purpose of Exercise Design

Design is the critical phase of creating an exercise activity. This is where you ensure that the activity will maximize learning value and cost-effectiveness, meet objectives, and support continuous capability improvements. The graphic in Figure 4-1 depicts the Design Phase.

The purpose of exercise design is to ensure your organization achieves maximum effectiveness from the exercise by

- ensuring critical issues will be addressed during conduct and considered from the beginning;
- maintaining focus on the objectives throughout the process, and therefore meeting costing expectations;
- laying the groundwork for, and in fact driving, development efforts in support of the exercise activity.

Overview of the Design Phase

There are nine components in designing an exercise activity. Although the results are different for tabletop exercises, drills, and exercises, the process is the same. The overall philosophy and rationale behind the design components are detailed below. We will then review the nine components first in the context of a tabletop exercise design and then in that of a drill or exercise. It is important to note that the nine components are interdependent considerations and not sequential in nature. Designing

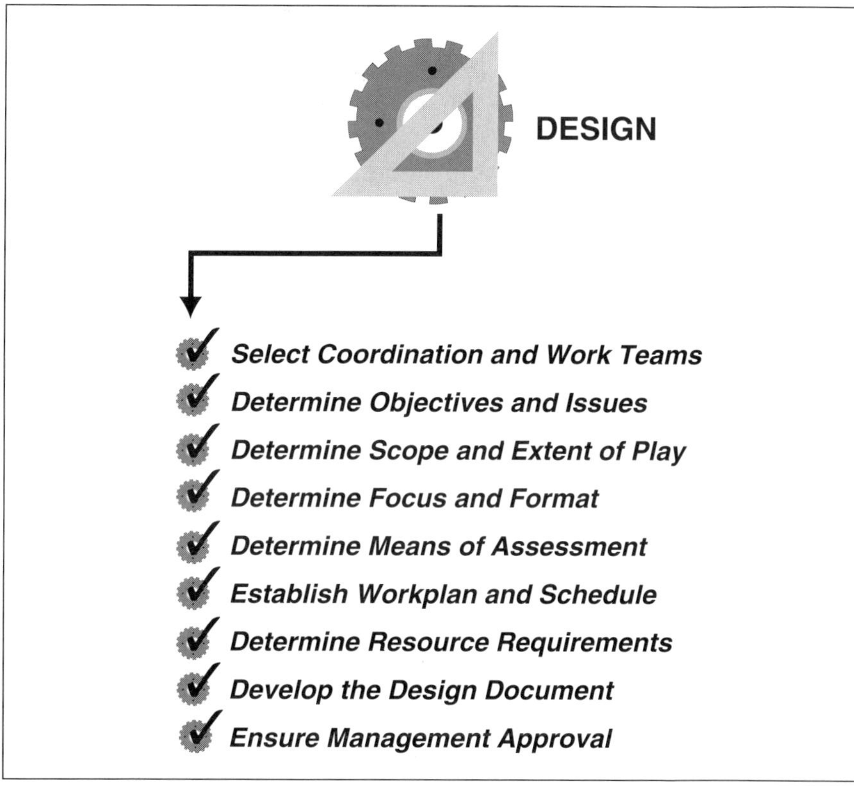

Figure 4-1
The Design Phase

any exercise activity is a dynamic process. First, the purpose of each component is described.

1. Select a coordinator and work teams: The coordinator is an individual charged with overall responsibility for the exercise activity. Work teams include personnel assigned to complete either an entire phase or a specific task within a phase in support of the exercise activity. As with other aspects of today's work environment, the use of work teams allows a more productive effort and enhances the quality of the final product.

It is important to select a coordinator at the very beginning of the process in order to assign clear responsibility for the exercise activity. Work teams are essential in completing the numerous tasks efficiently and with quality. If chosen skillfully, they also provide a diverse and solid core team for the entire exercise activity. The early designation of personnel allows an efficient and effective allocation of tasks and maximizes the use of available resources. Advance personnel planning also makes it easier for team members to allot the time and resources necessary to be an effective team member. Assign personnel with the appropriate knowledge level and capabilities to fulfill the requirements for the designated tasks.

2. Determine objectives and issues: *This is the most important component in the entire design and development process.* Objectives establish the purpose of the exercise activity and define what is expected to be achieved. Clearly stated objectives allow proper evaluation criteria to be determined and, therefore, success to be measured. Issues are identified vulnerabilities, capability gaps, unresolved policy or decision points, and important but perhaps sensitive items. Issues do not define the scope of the activity so much as they can dictate the type of scenario needed. This component defines *what* you want to achieve and *why*.

Establishing objectives and issues first is essential to lay the proper foundation for any successful exercise activity. These components determine the entire scope and extent of the exercise, whether a tabletop, a drill, or a full-scale exercise. They should guide your scenario development and other related actions, support material development, and administrative support requirements.

3. Determine scope and extent of play: The scope and extent of play establish the parameters within which your exercise activity will be conducted, including defining the duration, players involved, level of detail, and if appropriate, the level of simulation, extent of mobilization, and whether the date and time will be announced.

By containing all activities within these parameters, players are able to perform against established objectives, which provides a more effective learning experience.

4. Determine focus and format: The focus of any exercise activity is based on the extent to which training is to be accomplished versus the desired level of validation of capabilities. This decision is based on the identified issues and objectives. Although every exercise activity involves some level of both training and validation, it is important to determine the extent to which one is more relevant and appropriate to the teams involved than the other. Defining the format includes deciding on the type of exercise activity as well as the kind of format.

Formats for tabletop exercises may range from policy and planning development to response operations (process and procedures) and can involve multiple moves for a single scenario or multiple scenarios. Formats for drills and exercises range from limited or functional drills to full-scale exercises. To achieve maximum value from the exercise activity, each must be tailored to the capability, knowledge, and skill levels of the players. This is accomplished by determining both the focus and format during the design phase.

5. Determine means for assessment: This component involves establishing a way to evaluate the results of the exercise activity against the established performance standards and/or criteria. It identifies the audience and format for evaluation feedback.

Capturing and sharing the lessons learned during your exercise activity is vital to the continuous improvement of your crisis and emergency management capabilities. The method of assessment should be determined in advance to ensure sufficient information is gathered, documented, and analyzed during conduct; that proper evaluation is accomplished; and that the maximum sharing of key learnings is achieved.

6. Establish workplan and schedule: These are tools that guide your efforts through the design, development, conduct, and evaluation phases of exercising. They contribute to ensuring your exercise activities are accomplished through a managed process.

Regardless of the size or complexity of the activity, proper planning and adequate lead time for completion of activities is essential to maximize value and minimize inefficiencies. Therefore, it is important to establish a workplan and schedule of activities. With the greater complexity of drills and exercises, this component becomes even more important to ensure that all tasks are completed in a timely and efficient manner. Sample workplans for a tabletop exercise and for a drill or exercise appear later in this chapter.

7. Determine resource requirements: Manpower, facilities, and equipment resource requirements are needed for all exercise activities. These requirements are necessary to support a tabletop exercise, although it will require fewer resources than a drill or an exercise.

Making a preliminary assessment of resource requirements during this phase allows design personnel to better define their budget as well as reserve the necessary resources ahead of time. Again, the more complex the exercise activity, the more extensive your resources needs.

8. Develop a design document: A design document is the written documentation of your design process. By putting your decisions in writing, it facilitates a common understanding and agreement on the purpose and scope of the exercise activity. The design document is an effective tool to share with work teams to maintain focus. When provided to management, it outlines the planned activity and its goals and associated budget requirements.

9. Ensure management approval: Proper authority may have been given to appropriate personnel prior to the design of the exercise activity. If not, the coordinator should obtain approval by appropriate management of the design parameters of the exercise activity.

Now that you understand more about *what* needs to be done when designing an exercise activity, let's take a look at *how* you design the distinct types of activities. Remember, the design components are interdependent but do not necessarily represent a sequential process.

How To Design A Tabletop Exercise

Using the nine components, the following describes the unique design considerations of a tabletop exercise. Following this process will result in a solid design document, the foundation for all other related activities.

Component 1 - Select a tabletop exercise coordinator and work teams

Figure 4-2 outlines the basic criteria for selecting your tabletop exercise coordinator and work teams and provides tips to assist in making the proper selections.

Because tabletop exercises are not labor-intensive, one individual may play several roles. In fact, a small group of two or three people may be all that is necessary, depending on the size and number of groups involved.

Remember, an individual's value to a work team must be weighed against the value of his/her participation in the tabletop exercise. Do not include personnel who should actively play in the tabletop exercise.

Component 2 - Determine objectives and issues

Establish objectives that are achievable and challenging within the context of a tabletop exercise. They should also support the objectives set for the overall crisis management and exercise programming process.

Objectives and issues for tabletop exercises and discussion-based scenarios should be focused specifically on the team(s) or group(s) involved in play. They are determined based on

- the current stage of your group's crisis or emergency management capabilities;
- any identified vulnerabilities/gaps or issues for resolution for the team(s)/group(s);
- the level of knowledge and understanding of roles and responsibilities;
- applicable emerging issues within your organization, industry, or other area;

CRITERIA	TIPS
Select a **coordinator** who understands the exercise program, purpose of this activity, expected outcomes, and ideally some link or relationship to the team or group designated to play in the tabletop exercise.	The facilitator for conduct, or the designated evaluator, typically is best-suited to fulfill the coordinator role.
Select personnel for the **design** team who understand the process and/or the needs of the groups who will play in the tabletop exercise.	Both the facilitator and evaluator should serve on the design team. Also consider involving any lead technical or operating location trusted agent.
Select personnel for the **development** work teams who have both a technical and broad understanding of how the incident would evolve and who can identify the critical issues the group should address.	A lead developer should be designated and is usually best suited to be the evaluator. Identify the full extent of technical and "external world" trusted agents, regardless of level or extent of involvement needed. The facilitator must be on the development team.
Select members for the **logistics/administrative** work team who have experience and access to the type of resources required for a tabletop exercise.	A tabletop exercise is typically conducted in a conference room setting. Likely resources include a conference room large enough to accommodate the team(s) and observers, flip charts, and possibly refreshments.
Select personnel for the **conduct and evaluation** team who are experts in facilitation and evaluation.	The facilitator should have expert facilitation skills and not be an integral player of the team(s) involved or a normally critical resource to those participating.

Figure 4-2
Tabletop Exercise Personnel Selection Criteria

- knowledge and understanding of capabilities requiring validation.

Component 3 - Determine scope and extent of play

To determine scope and extent of play, you must first determine the identity of the players based on the group's needs in addition to the level of knowledge and objectives previously established. You will also need to determine how long the tabletop exercise will last and how detailed to make the scenarios, based on the objectives and number of issues you want to address. Tabletop exercises may be designed to support a single team or a group of teams and typically involve only discussion.

Component 4 - Determine focus and format

Selecting the focus and format for a tabletop exercise is accomplished by considering the following.

1. training vs. validation
   - For example, if personnel are just beginning to develop crisis and emergency management capabilities, a greater emphasis should be placed on training. As the process is established and personnel further develop their capabilities, the focus of tabletop exercises should move toward validating knowledge.
   - The tabletop exercise format can be a vehicle for both training and validating capabilities for specific issues and objectives.

2. policy vs. process tabletop exercises
   - Tabletop exercises can be conducted in a range of formats, from policy or development to process validation. The extent to which training versus validation is the primary focus will also suggest the appropriate type of tabletop exercise. Selection of the one best suited to your objectives and level of capabilities is essential.

3. multiple scenarios vs. multiple moves
   - The multiple scenario format is useful in introducing players to a wide range of incidents from which they can compare and contrast response activities. This approach is often used during the development stages of a crisis and emergency management capability but can be useful at any stage. The multiple scenario format is also useful for tabletop exercises involving numerous response groups.
   - A single scenario with multiple moves is optimal when discussing phases of an operation or different response

activities occurring within a time sequence. This format tends to be utilized most to validate and reinforce process. It works well with both single and multiple teams.

Component 5 - Determine means of assessment

As with any exercise activity, it is critical to identify the means of evaluation early in the design process to ensure proper assessment and feedback is possible. When determining the means to assess a tabletop exercise, be sure to identify how you will

1. capture the results,
    - The types of information and results expected from the tabletop exercise should be anticipated. Player and facilitator evaluations are essential to capture, and assigning a specific evaluator to document discussions, decisions, and consensus of the group works well. You should also include an action planning session at the end of the tabletop exercise. This is to be conducted by the facilitator and summarizes the players' key learnings and proposed action items. Results are frequently captured through the use of evaluation or critique forms.

2. analyze the results and measure success,
    - How performance will be measured is usually addressed by comparing discussions, proposed actions and decisions, and demonstrated knowledge against objectives and supporting evaluations criteria. Assess the results of the tabletop exercise discussions by reviewing both the verbal and written documentation.

3. present assessment findings,
    - The format can vary from formal written reports to informal verbal feedback. The level of detail can also vary from detailed to a general overview.
    - Given the interested and defined audience, determine the appropriate method and distribution for the evaluation findings.

4. utilize findings to improve your organization's capability.
    - Consider how you will maximize the value of the exercise by sharing lessons learned throughout your organization. This may be accomplished by including others in your organization or from other locations as observers in the tabletop exercise. This can also be accomplished by expanding the distribution of evaluation information.

- Use a system to track action items and monitor improvements.

Component 6 - Establish workplan and schedule

Your tabletop exercise workplan and schedule provide the road map for complete and efficient work efforts. Although on a much shorter time frame than those of drills or exercises, these tools should

1. establish milestone and completion dates ensuring ample coordination and scheduling time;
2. allow adequate time for technical development, material reviews, and appropriate approvals;
3. assign tasks to appropriate personnel;
4. encompass all the components of the entire tabletop exercise design, development, conduct, and evaluation process.

A sample tabletop exercise workplan is contained in Figure 4-3.

Component 7 - Determine resource requirements

Resource requirements for tabletop exercises are usually minimal. However, there are several areas that should be considered each time, including reviewing

- the scope and objective,
- duration and format of the tabletop exercise,
- number of personnel to play and those expected to observe.

Make sure to consider

- personnel resources to develop the scenario, as well as facilitate, evaluate, develop and present findings, and track lessons learned;
- facilities for conduct and debriefings;
- refreshments;
- equipment and supplies such as overhead projectors, VHS playback deck and monitor, notepads, flip charts or white/copy boards, etc.;
- resources needed during each phase, from design through evaluation.

Component 8 - Create your design document

Develop your design document, recording all decisions on the objectives, scoping, and workplan.

Component 9 - Ensure management approval

Using the design document, the coordinator should discuss with appropriate management the

Instructions: *Assign actual completion dates in the space provided by each week.*

Design Phase

Week 1 - _____
- *Determine focus of tabletop exercise, i.e., appropriate mix of policy or operational orientation, and training/validation focus.*
- *Determine objectives.*
- *Define internal requirements for budgets and approval process.*
- *Determine number of scenarios.*
- *Determine composition of scenario development and conduct team and assign responsibilities.*
- *Determine format and method of assessment to be used; e.g., report, presentation, etc.*
- *Develop detailed milestones for completion.*

Development Phase

Week 2 - _____
- *Research incident(s).*
- *Decide on type and severity of incident(s).*
- *Draft scenario outline(s) and events timeline.*

Week 3 - _____
- *Review outline, agree on number and content of moves (if applicable), and extent of detail to be included.*
- *Draft annotated outline.*
- *Review and approve annotated outline.*
- *Begin content development.*

Week 4 - _____
- *Review first draft of scenario.*
- *Review scenario based on team components.*
- *Determine what other participant materials are needed and begin development of materials.*
- *Determine what facilitation aids will be required and begin development of materials.*

Week 5 - _____
- *Review second draft of scenario and revise.*
- *Review final draft of scenario and participant and facilitator materials.*
- *Incorporate comments and finalize scenario, and revise participant and facilitator materials.*

Week 6 - _____ (prior to conduct)
- *Final review and approval of scenario, participant and facilitator materials, and other materials.*
- *Dry run for conduct.*
- *Begin production of scenario(s) and player and facilitator materials.*
- *Put together scenario packages for participants and facilitators.*

Conduct Phase

Week 6 - _____ (day of conduct)
- *Conduct the tabletop exercise.*
- *Evaluate performance and assess training value.*
- *Conduct "hotwash" debriefing session with players and facilitators.*

Week 6 - _____ (day after conduct)
- *Debrief among facilitators and evaluators.*
- *Assess design, development, conduct, and evaluation processes.*

Evaluation Phase

Week 7 - _____
- *Review participant comments.*
- *Develop outline and begin drafting evaluation for report or presentation purposes.*
- *Formulate recommendations for modifications to policy, program, plan, or procedures.*
- *Complete draft of report/presentation.*

Week 8 - _____
- *Review evaluation report/presentation among facilitators and evaluators.*

Week 9 - _____
- *Incorporate comments and revise final version.*
- *Present findings to appropriate managers and personnel.*

Figure 4-3

Model Tabletop Exercise Workplan

- budget associated with the tabletop exercise, given the scope, duration, and format;
- personnel/support requirements;
- potential policy, program, or developmental implications likely to be highlighted.

These discussions are also a good opportunity to solicit support and commitment from management to the exercise program and the overall crisis and emergency program.

How To Design A Drill Or Exercise

The design of a drill or exercise follows the same basic process as that of a tabletop exercise. However, the effort is far greater and more complex in nature. Although drills and exercises vary in focus, scope, and complexity, they are addressed together as their design process and considerations are largely the same. The following takes you through each step in the design for a drill or exercise, again with the task of creating a design document in mind.

Component 1 - Select a drill/exercise coordinator and work teams

Working groups should be established to support the requirements of each of the four phases of exercising activity (design, development, conduct, and evaluation). Assign personnel with the appropriate knowledge level and capabilities to complete the designated tasks, while ensuring they will not be critical as players to the response effort being validated. Due to the greater personnel resource needs of a drill or exercise, it is essential that work team members are screened and selected carefully in order not to undermine conduct. Figure 4-4 provides selection criteria and tips for staffing the coordinator and work team positions to support the design, development, conduct, and evaluation of your drill or exercise.

Establish the size and composition of each group by considering the

- type and size of drill or exercise;
- scope and extent of play, including duration;
- number of response locations and simulation requirements; and
- experience of personnel selected.

Personnel can be part of more than one working group, though the larger the exercise, the less likely this is. The more experience a person has, the more roles he or she may be able to support. Remember, an individual's value to a work team must be weighed against the value of his or her participation as a player in the drill or exercise.

CRITERIA	TIPS
Select a **coordinator** who is experienced in drills and exercises and clearly understands the objectives and scope requirements. Having effective project management skills would be desirable, given the scale, complexity, number of tasks, and personnel involved.	The person selected to function as the coordinator should be knowledgeable of the response operations likely to be involved in play. Typically, the person responsible for the crisis or emergency management program at that level is the best candidate. Exercising experience is a must.
Select personnel for the **design** team who understand the response team and design process(es) and/or the unique needs of the groups who will play in the drill or exercise. This core design team will usually guide the entire drill and exercise through completion.	The design team members are typically those with some responsibility or accountability for various aspects of the organization's crisis and emergency management program. A lead Trusted Agent from each technical area or organizational level is usually selected. A group of four to seven carefully selected individuals is best.
Select personnel for the **development** work teams who have both a technical and a broad understanding of how the incident would evolve, who understand the external factors that can drive and impact upon the situation and organization, and who can identify the critical issues the playing group(s) would address if the event were real. It is also helpful to assign personnel with special skills to other developmental items.	The development team is usually subdivided into a core development team responsible for overall scenario coordination and integration as well as the narrative and timeline development. Individual functions or specialty areas, such as the news media, legislatures, or non-playing facilities will require one or more trusted agents to develop appropriate scenario aspects and specific messages. The total number of development team trusted agents can range from 5 to 15, with varying levels of responsibility and involvement. The core design team is always part of the development team(s). Development trusted agents are almost always assigned simulation or other conduct responsibilities for the conduct phase.
Select members of the **logistics/administrative** work team who have experience and access to the many and varied resources that may be required.	Logistics and administrative support to carry off a drill or exercise can be extensive, from conduct facilities, communications and computer equipment, to logistics arrangements to support multiple shifts and/or night operations. Based on the requirements, team members should be experienced and have appropriate access and/or authority to procure needed resources, make appropriate arrangements, and install or set up for conduct operations. This team can include between 1 to 7 members.
Select personnel for the **training/briefing** work team who have instructional expertise and who are familiar with each of the conduct roles, including controllers, evaluators, and simulators.	Personnel selected to support the training/briefing efforts usually include the coordinator, the simulation cell lead(s), and one or two others who are likely to be members of the core design team. Staff to support development needs and production (slides, overheads, handouts, etc.) should also be identified.
Select personnel for the **conduct** team who are experts in simulating (role-playing certain groups or individuals) or controlling exercise play, or who are knowledgeable enough about each response group, as well as effective response management, to evaluate designated response locations.	The conduct team is described in more detail in Chapter 5 - Development. The coordinator oversees the entire conduct process, and either key or all members of other work teams can be assigned conduct roles. Supplemental personnel should be selected based on the role to be performed or the function or area assignment. The conduct team can range from 7 to 120+ members, depending on the extent of the activity and the need for shifts.
Select personnel for the **evaluation** team who understand the essential aspects of performance evaluation and who can design, develop, and execute an effective evaluation plan.	The evaluation team should be comprised of at least one qualified (independent, peer, and/or third party) individual per response location, in addition to an overall evaluator lead (from the core design team). The evaluation team can range from 2 to 40 members.

Figure 4-4
Drill/Exercise Personnel Selection Criteria

Component 2 - Determine objective and issues

The larger the scale of the exercise activity, the wider the range of issues and objectives you can address.

Establish objectives that are achievable and challenging for the players. These should also support the objectives set for the overall crisis and emergency management program as well as your exercising program. The objectives should be oriented toward the actual demonstration and validation of capabilities.

Objectives and issues should be determined based on

- identified vulnerabilities or gaps;
- the current stage of your group's crisis management program;
- capabilities and response concepts requiring validation, including functions and integration aspect;
- the level of knowledge and understanding of roles and responsibilities by both individuals and teams;
- regulations driving your planned response;
- emerging issues.

Component 3 - Determine scope and extent of play

Based on the objectives, the following should be addressed when defining your drill or exercise scope and extent of play:

- Who will play (organizations, teams, individuals)?
- What level of simulation is required to maintain realism (non-participating external and internal groups)?
- To what degree will personnel and/or resources be mobilized?
- What should the duration of the exercise activity be in order to allow all objectives to be addressed?
- Do objectives suggest an appropriate date, time, and/or location(s) of the drill or exercise?
- Will the drill or exercise be announced or unannounced?

Component 4 - Determine focus and format

Selecting the focus and format of a drill or exercise is accomplished by considering

1. training vs. validation,
    - For example, if personnel are in the early stages of developing crisis management capabilities, a greater emphasis should be placed on response role training. As personnel and teams develop their capabilities, the focus should move toward validating both skills and knowledge.
    - The scope and format of drills and exercises lend themselves to validating capabilities, although they also provide one of the best opportunities for training.

2. range of drills through exercise,
    - The specific objectives and issues, and the extent of play, scale, and scope will suggest an appropriate type.

3. number of response locations expected to be involved.

Remember, the activity format can range from a limited drill to a full-scale exercise. Make sure the focus and format you select matches the objectives, scope, and scale anticipated. This will help ensure appropriate expectations and ensure requirements are established.

Component 5 - Determine means of assessment

Due to the validation nature of drills and exercises, an effective means of assessing performance must be planned from the design phase. A drill or exercise is more complex and can involve from four or five to thousands of players. The ability to observe response actions, effectively document and assess performance, and provide appropriate feedback to these players requires planning, appropriate tools and materials, and a defined system. When planning your evaluation, you should consider the way in which you want your team to:

1. capture the results,
    - From the objectives, issues, designated response teams, and locations, identify the capabilities being validated in the drill or exercise. Identify the information and types of documentation required to support effective evaluation. Evaluation forms are frequently created to ensure proper documentation.
    - Design appropriate tools to allow proper collection and processing of observations and information. Evaluation is usually supported with specially designed evaluation forms.

2. analyze the results and measure success,
    - How performance will be measured is usually based on comparing response actions and decisions to plans and procedures, effective crisis and emergency management techniques, and the desired or expected performance level. Evaluators have the primary responsibility for this analysis and should be carefully selected and assigned to achieve proper evaluation.
    - Ensure you plan to provide ample opportunity to receive verbal as well as written feedback from the players, controllers, and simulators. Having been directly involved in the drill or exercise, their insight and personal critique will be invaluable. Scheduled debriefing sessions and critique forms will provide you with their analyses.

3. present assessment findings,

- The format, level of detail, and audience(s) should be identified early in the design phase. Evaluation efforts must be accomplished throughout development, conduct, and the evaluation phase, and these efforts will be driven by the design foundation. Consider
- Written/formal vs. verbal/informal
- Detailed vs. general highlights/overview
- Determine the best means to distribute the findings based on the audiences.

4. use findings to improve your organization's capability
    - Capture lessons learned
    - Identify action items
    - Track action items

Component 6 - Establish workplan and schedule

Your drill or exercise workplan and schedule will be critical to ensuring the activity is valuable, of a high quality, and managed and carried out cost effectively and efficiently. The workplan should

1. identify established milestone and completion dates, ensuring ample coordination and scheduling time;
2. allow adequate time for internal and, if appropriate, external review and appropriate approvals of all plans and materials;
3. assign tasks to personnel, as appropriate;
4. encompass all the nine components of the four phases of an exercise activity (design, development, conduct, and evaluation).

A sample Drill or Exercise Workplan is contained in Figure 4-5.

Component 7 - Determine resource requirements

Determine resource requirements by reviewing

- the scope and objectives,
- duration of exercise,
- number of personnel,
- special simulation needs.

Make sure to consider

- functional facilities to support the control and simulator organizations;
- equipment and supply needs at all locations and by all those in the conduct organization, such as communications equipment for controllers, phones and/or radios for simulators, etc.;
- number of simulation cells required;
- refreshments;
- resources needed during each phase of the drill or exercise, not just during the actual conduct.

Instructions: *Assign actual completion dates in the space provided by each week.*

Design Phase

Week 1 - _____ • Outline project status, objectives, trusted agent needs, and participants.

Week 2 - _____ • Refine objectives, determine exercise size/scale, designate exercise development team, develop detailed milestones for completion of project.

Week 4 - _____ • Determine method and format of evaluation process.

Week 5 - _____ • Finalize objectives, participants, degree of simulation versus actual demonstration, assign exercise development team members assignment.

Development Phase

Week 6 - _____ • Research possible scenario options.
• Begin outlining logistical needs for exercise.

Week 7 - _____ • Select type, severity, and location of incident.
• Begin defining exercise organizational structure.

Week 8 - _____ • Draft scenario outline and events timeline.

Week 9 - _____ • Assess level, type, and duration of training for the exercise organization.
• Continue designing exercise organizational structure including controller lead and team, evaluator lead and team, and simulator lead and team.

Week 10 - _____ • Review and revise scenario outline and events timeline.
• Develop training plan for each group and begin development of training and briefing materials.

Week 11 - _____ • Obtain exercise coordinator approval on scenario.
• Contact trusted agents to confirm availability and to discuss confidentiality guidelines, their role, and the time commitments.

Week 12 to Week 13 - _____ • Begin scenario content development and first draft of narrative.
• Send scenario outline to trusted agents to review.

Week 14 to Week 15 - _____ • Review revised events timeline and first draft of narrative.
• Identify/discuss functional areas, exercise locations, and issues requiring messages to prompt activity.

Week 16 to Week 17 - _____ • Revise narrative based on team discussions.
• Determine space requirements of control cell, general exercise operations and coordination, conference areas, training areas, etc.

Week 18 - _____ • Review second draft of narrative and discuss issues to be addressed by each functional area.

Week 19 - _____ • Complete development of events timeline and narrative.

Week 20 to Week 24 - _____ • Begin message development.

Week 25 to Week 26 - _____ • Continue message development and begin reviewing messages already created for content.

Week 27 to • Determine hardware and supply requirements for exercise organization.

Week 29 - _____ • Determine appropriate communication system.

Figure 4-5
Model Drill/Exercise Workplan

Week 30 - _____	*• Determine what materials are needed to support conduct and assessment of exercise.* *• Complete message development.*
Week 31 - _____	*• Conduct an internal run-through of scenario and messages.*
Week 32 - _____	*• Revise messages as required and complete development of any additional messages.*
Week 33 - _____	*• Produce any training and/or briefing materials.* *• Produce any materials needed to support conduct and assessment of the exercise.*
Week 34 - _____	*• Determine refreshment needs over the course of exercise.* *• Finalize all logistic arrangements.* *• Determine security considerations and make appropriate access and other arrangements.* *• Package and ship pretraining materials to exercise organization.*
Week 35 - _____	*• Conduct player briefings and provide player groundrules.*
Conduct Phase	
Week 36 - _____ (prior to conduct)	*• Conduct training/briefings of exercise control leads.* *• Conduct controller/evaluator/simulator training/briefings.* *• Conduct dry run of simulators.*
Week 36 - _____	*• Conduct exercise with hotwash sessions and player debriefings immediately afterwards.*
Evaluation Phase	
Week 37 - _____	*• Review evaluator reports.* *• Review player comments.* *• Develop outline of evaluation report.*
Week 38 - _____	*• Begin drafting evaluation report.*
Week 39 - _____	*• Review evaluation report.*
Week 40 - _____	*• Incorporate comments and revise.*
Week 41 - _____	*• Complete draft of evaluation report.*
Week 42 - _____	*• Finalize report.* *• Develop presentation materials.* *• Determine distribution.*
Week 43 - _____	*• Produce report.* *• Present findings of report to players and management.*
Week 44 - _____	*• Action planning meeting and exercise review.* *• Distribute reports as appropriate.*

Figure 4-5
Model Drill/Exercise Workplan (cont.)

Component 8 - Create your design document

Develop your design document, recording all decisions on the objectives, scope, and extent of play. The format should be defined as well, and the means of assessment outlined. The workplan should be included in the design document. This document will become a primary tool to guide the coordinator and work teams through the extensive development, conduct, and evaluation efforts for a drill or exercise.

Component 9 - Ensure management approval

Using the design document as the basis, the coordinator should discuss with appropriate management the:

- budget associated with the drill or exercise, based on the scope, number of players, locations, and duration;
- development/conduct/evaluation personnel, equipment,supplies, and support requirements;
- policy, program, and capability implications likely to be highlighted;
- the importance of management support and commitment.

These steps of design are essential for all involved to achieve a successful exercise activity. Design is the foundation upon which all development conduct and evaluation activities are based.

HOW TO

Create a Design Document

- ☐ *Select Your Coordinator and Workteams*
- ☐ *Identify Objectives and Issues*
- ☐ *Determine Scope and Extent of Play*
- ☐ *Determine Focus and Format*
  - *- Training and validation*
  - *- Range and format type of exercise activity*
- ☐ *Establish Workplan and Schedule*
- ☐ *Identify Resource Requirements and Budget*
- ☐ *Ensure Appropriate Management Approval*

CHAPTER 5

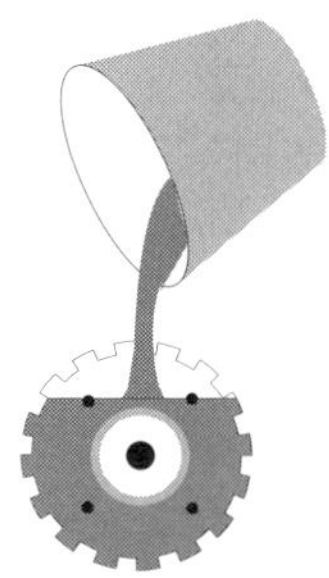

Achieving Realism With Development

The most useful simulations are those that exercise well-developed management systems as part of an intregrated approach to crisis preparedness. They must relate directly to clearly identified risks, emphasize substance over style, and offer a meaningful professional challenge to participants. Perhaps most important, simulations should be designed, conducted, and evaluated not as "pass-fail" tests, but as a means of assessing and improving an organization's collective skills.

Mary Woodell
Senior Principal
CMRC & Co.

This chapter takes you through the steps of development in order to create, fully prepare, and organize an exercise activity.

The objectives are to help you

- understand the components of the development phase,
- develop your own exercise activity.

Introduction

Using the design document as the foundation, you are now ready to begin the development phase. Many of the decisions made on the design of your activity will dictate the development approach and requirements.

Purpose of the Development Phase

It is during the development phase that the steps are taken to create, fully prepare, and organize the exercise activity.

Overview of the Five Components of the Development Process

The process to develop exercise activities can be divided into five primary components. These include

1. scenario development
2. support materials development
3. training and/or briefing session development
4. logistics and administration
5. final production

Many of these activities are interrelated and will be undertaken concurrently. Therefore, they are *not* to be viewed as sequential steps. The graphic in Figure 5-1 depicts the development phase.

Due to the differences in size and scope between a tabletop exercise and a drill or exercise, the development phase for each exercise activity will be vastly different. Although you will work through the same five components, the actual tasks within these areas are distinct and will be discussed in detail by type of exercise activity. First, the purpose of each component is described.

Component 1 - Scenario development

The scenario is the foundation of an effective exercise activity and lays the groundwork for accomplishment of established objectives.

Selection of an appropriate scenario should be based on the objectives and such considerations as the vulnerabilities, hazards, or weaknesses

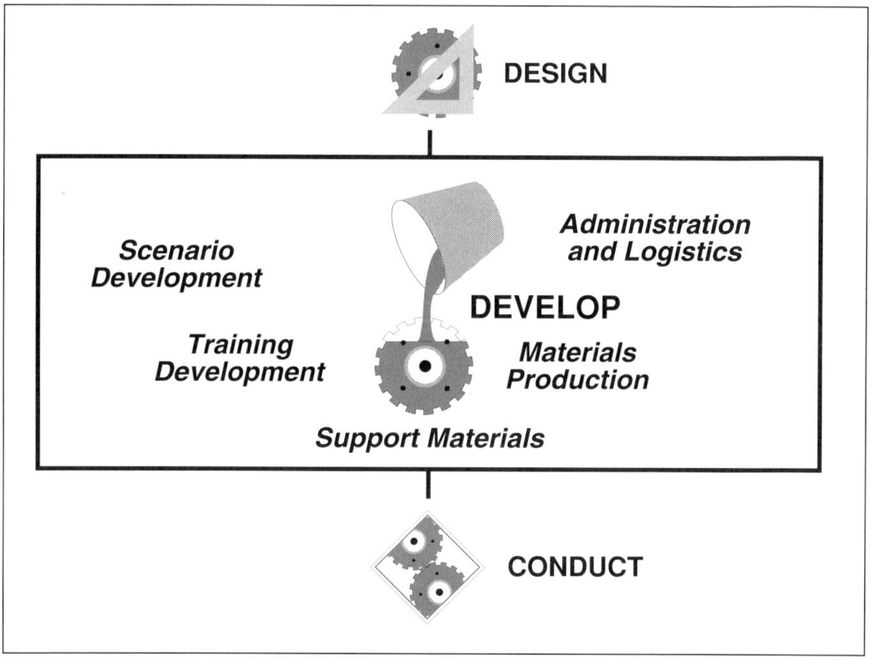

Figure 5-1
The Development Phase

that are of concern to the organization specifically or to the industry as a whole, and the probability there is of occurrence.

Regardless of whether you are developing a tabletop exercise, drill, or exercise, there are some general tasks you will need to complete to accomplish scenario development. These include

- selecting trusted agents with technical and/or functional expertise to provide guidance and input into the scenario;
- gathering the background information needed to make your scenario realistic, through trusted agents or external research;
- utilizing or accessing data modeling or computer scenario generating capabilities, if applicable and available;
- selecting additional personnel to support completion of the scenario development process.

General process for developing a scenario

There are five basic steps to developing a scenario. These steps include

1. research, including technical and external information;
2. drafting and integration;
3. review;
4. talk-through;
5. finalization.

The research step begins during the early stages of design and development, while assessing scenario and location options before selecting the final scenario. From there, the research conducted is of a technical nature and externally focused. The technical accuracy, as well as the realism of the external forces that would come to bear in a real situation, are critical to the success of the exercise activity and your program overall. The momentum and commitment to the crisis and emergency management program can directly benefit from the realism of your exercise activities.

While drafting the scenario is the primary scenario development activity, the integration and periodic review of what may be several efforts will ensure a cohesive scenario.

When finally put together, a talk-through is a useful technique to ensure the scenario events:

- flow,
- have the proper timing,
- are integrated,
- are realistic,
- are complete.

If a talk-through is not done, a designated trusted agent should review the total scenario package for accuracy and completeness. At this point, the scenario can be finalized. Comments and discrepancies discovered during the review and talk-through should be incorporated, and the scenario development completed.

Component 2 - Develop support material

Support materials provide important context for the scenario and should be developed in tandem with the scenario. The type of support materials for a tabletop exercise tend to differ dramatically from those for a drill or exercise. However, the kinds of support materials can generally be categorized as player and/or conduct organization forms, player ground rules, and other supporting materials for effective conduct. More detailed descriptions of specific supporting materials for tabletops and drills/exercises are outlined in the unique sections following.

Component 3 - Develop training or briefing session

An essential part of a successful exercise activity is the ability of the exercise organization and players to perform their roles effectively. It is important that there be a common understanding of the objectives, ground rules, limits of play, and responsibilities of each group. Pre-exercise briefings and/or training are important tools for ensuring that everyone benefits from every exercise activity.

When developing a training or briefing session, you should identify the

- groups involved (e.g., players, controllers, evaluators, simulators, observers),
- format (training and/or briefing),
- types of supporting material (overheads, slides, handouts, handbooks),
- tentative schedule for the session(s).

Component 4 - Develop and arrange logistics

Developing and arranging logistics actually begins in the design phase with the listing of initial logistic needs (i.e., facility setup, supplies, meal planning, etc.) and ends at the completion of conduct with the breakdown of the facilities, return of equipment, and collection of any remaining exercise activity materials. It is during the development phase, though, that most logistic arrangements are completed.

Component 5 - Develop the final "product"

This final stage of development encompasses the completion of pre-conduct preparation and production activities. The preparation of high-quality materials in support of training/briefing sessions and the exercise activity itself is critical to smooth operations during play. While preparing to produce materials, ease of use by players should always be considered for the packaging.

Once your materials have been developed, reviewed for quality, produced, and delivered to the appropriate personnel, you are ready for the main event. The next step in the exercising process is conduct.

How To Develop *A Tabletop Exercise*

Component 1 - Develop the scenario

Research: Though the scope and length of the tabletop exercise scenario may be limited compared to larger scale drills or exercises, you will still need to research the technical and external aspects of the scenario to ensure realism. Consider the research needs based on the technical nature of the postulated event, your organization's expected response, and the external stakeholders and others who might play a role if the event were real. Much more information will be obtained than is actually written into the final tabletop exercise scenario. Some of the information will be included in the facilitator notes discussed later.

Drafting and Integration: Once you have gathered the necessary background information, you need to decide how you are going to present the information. You may wish to use either a single scenario or multiple scenarios, depending on the objectives.

A single scenario with one or more moves is optimal when discussing phases of operations or when different response actions occur during a running time sequence. Multiple scenarios are useful to introduce players to a wide range of incidents to which they can contrast and compare response activities. A background piece is also typically developed for tabletop exercises. This piece provides the context for the postulated event, i.e., it sets the stage.

Once the format is determined and the research is underway, you can begin to draft and integrate the scenario information into the final product.

The different components for development of a single scenario include

- Background
- Scenario Move 1
- Scenario Move 2
- Scenario Move 3

Each scenario move provides details of events that have transpired and response actions that have been taken during a specific time frame. Each move is time sequenced and introduces the scenario events in chronological order. The following are sample scenario background piece and two scenario moves.

Sample

Metropolitan Produce Tabletop Scenario

Background

It is mid-May and the asparagus harvest is at its height. This year's crop is plentiful despite early problems with an unusual fungicide-resistant strain of a common asparagus fungus. Metropolitan Produce, the largest wholesale produce distributor in the northeast, has operations in New York, New Jersey, and southern New England. On Monday evening, May 13, Metropolitan's Brooklyn warehouse received a large shipment of asparagus from one of the region's largest commercial farming operations. Metropolitan workers stored the shipment overnight in the Number 3 refrigeration unit.

Early Tuesday morning, May 14, workers opened the Number 3 refrigeration unit to begin loading the asparagus onto distribution trucks. Upon doing so, they found the unit's cooling equipment had failed sometime during the previous night and the temperature inside was approximately 12 degrees warmer than it should have been. A large amount of what appeared to be condensation had collected on the boxes and exposed produce, but the temperature increase did not appear to have a negative impact on the product. In accordance with company policy, the workers notified the warehouse manager of the incident.

After conducting his own inspection, the manager determined that the produce was unaffected by the temperature rise and approved shipment of the asparagus. He then returned to his office and phoned Metropolitan headquarters with a request for refrigeration maintenance personnel. The workers began loading the asparagus onto distribution trucks which delivered the produce to Metropolitan's regular restaurant and retail outlet customers.

When maintenance personnel arrived, they discovered a loss of pressure in the coils that circulate coolant through the refrigeration unit. They also determined that a considerable portion of the coolant leaked from the coils. They advised the warehouse manager that the coolant could have leaked onto the produce stored in the refrigeration unit.

Move One

It is 10:00 a.m. on Wednesday, May 15. Metropolitan headquarters in Manhattan has received the following information:

- The Brooklyn warehouse has received several phone calls in the last 36 hours from restaurants complaining of an acrid taste to the asparagus they received on Tuesday. Several of the restaurants have called a second time to relay complaints from diners who say they are suffering from food poisoning.
- At 8:30 this morning, a woman called the Consumer Correspondence in New York to complain about some asparagus she had purchased at a grocery store the day before. She complained that it had an unusually bitter taste and that her two children awoke before dawn this morning complaining of stomach pains. She was concerned that her children's illness was connected with a chemical insecticide that may have been sprayed on the asparagus in the field.
- A Metropolitan salesperson phoned company headquarters to convey complaints he has received concerning the asparagus

shipped to his retail customers the day before.

- A New York Post reporter has contacted headquarters to confirm a rumor about tainted produce.

Move Two

It is four hours later.

- Electronic media have picked up the story and are including related stories in afternoon news broadcasts. The stories are speculating on the cause of the problem and have raised the possibility of everything from acid rain to pesticide poisoning. Metropolitan headquarters is being hit with a flurry of calls over the incident.
- Among the calls are several from hospitals that have admitted patients complaining of food poisoning. Several are suffering from severe vomiting and nausea.
- The New York Department of Health Services is running tests on samples it has collected. At this point, no results are available.
- Several of Metropolitan's largest customers are insisting Metropolitan replace all unused produce and cover their costs. A few are threatening to transfer their business to other produce distributors.

(This is a hypothetical situation for exercise purposes only.)

The different components for development of a tabletop exercise using multiple scenarios are

- Background
- Scenario #1 Snapshot
- Scenario #2 Snapshot
- Scenario #3 Snapshot

A scenario snapshot is a single view of an incident that provides relevant background information and an overview of the events and response underway. It is not usually time sequenced, and like a photo provides a "snapshot" of the scenario. The following is a sample scenario snapshot.

Sample

Tabletop Scenario #1

Current Scenario

At 10:00 a.m. today, a receptionist at Bank X's sixth floor executive offices received an anonymous call on an outside line. The unidentified caller said that "the sixth floor executives are about to get what they deserve. Bombs are set to go off at any minute." In accordance with company procedures, the receptionist tried to obtain the identity of the caller, but he hung up immediately.

The receptionist immediately reported the bomb threat to security, and security notified John Smith in Administrative Services. At approximately 10:20, an explosion occurred near the Trading Room on the second floor. An impromptu evacuation had begun but was not yet completed when the bomb detonated. No one, however, was injured. The explosion itself caused minimal structural damage, but the resultant fire has rendered the entire area inaccessible. Equipment and data damage are likely to be severe from both the flames and the water used to extinguish them.

A bomb squad is searching the entire building, including the garage. The squad located a second bomb considerably more powerful than the one that already exploded. The squad is trying to disarm the second bomb.

Reporters have arrived at the scene and are not only filming the fire but also are interviewing Bank X employees, many of whom are noticeably shaken.

(This is a hypothetical situation for exercise purposes only.)

Review: After you have drafted all the different parts of the scenario, you will want to submit it to the trusted agent(s) and the selected facilitator(s) for review input and feedback. The trusted agents can check that the scenario is technically correct and identify whether any relevant issues have been missed. The facilitator can check the scenario in relation to meeting the tabletop exercise objectives established in the design phase. As a note, it is important when constructing your workplan to schedule enough time for the review. Some of the most valuable insight into the scenario will be given during this step.

Talk-through: A talk-through is an informal practice run, which is optional for a tabletop exercise. It can simply consist of simply a final review by the facilitator(s) to ensure he or she is familiar with the scenario, facilitation points, and essential issues to be raised. However, a discussion should occur between either the facilitator and evaluator or between the trusted agents and lead scenario developer to ensure the scenario is complete and properly focused.

Finalize: This step involves incorporating into the final scenario all comments and additional input received during the review and talk-through steps.

Component 2 - Develop support materials

Support materials for a tabletop exercise include

- Ground rules of play

Ground rules of play are the instructions for the tabletop exercise given to the players. They also include helpful tips so that players can get the most out of their participation in the tabletop exercise. The following is an example of a tabletop exercise ground rules piece for player.

Sample

Tabletop Exercise

Ground Rules And Tips For Participants

You will be participating in a tabletop exercise that will provide you with an opportunity to consider the policies, plan, and procedures your organization currently has or should have in place. Our objective today is to clarify your understanding of roles and responsibilities through facilitated discussion. You will not be required to perform actual response duties, but you will discuss what actions or decisions you would make when confronted with different crisis situations.

Three scenarios will depict independent hypothetical incidents involving your organization's operations. Each of the scenarios has significant potential to impact one or more operating areas. As you review these scenarios, you should consider what actions are appropriate based on the background information provided, as well as actual events and current economic and political conditions.

After reviewing each scenario, we will take approximately 10-15 minutes to consider the appropriate decisions and courses of action which would be required. A series of discussion questions will be provided to assist you to focus your thoughts on the implications of the information provided. Following this, we will facilitate a general discussion of your analysis and decisions. This discussion will permit the exploration of alternative policies, decisions and actions, as well as an examination of key issues relevant to the incident.

To help you get the most out of this experience, we recommend that you keep the following lists as we work through the scenarios:

- issues and points where you would like additional clarification
- specific planning or procedural improvements or additions
- training needs identified through discussion
- resource requirements to implement identified needs

There will be an opportunity at the end of the session to consolidate and prioritize what players view as the next steps for your company to enhance its crisis management capabilities. The lists you develop during the session will be useful in identifying next steps for the overall program and for individuals with crisis management responsibilities.

Treat the scenarios as real events. Participate openly. The more you ask questions and present your thoughts candidly, the better this exercise will be. You should keep today's objectives in mind throughout the exercise.

Question Forms: Question forms aid the players in collecting and focusing their thoughts when reviewing the scenario. Question forms are also a way for players to express ideas or concerns that they may not feel comfortable expressing during the general discussion. A sample of a tabletop question form that would be used with a scenario follows.

Sample

Tabletop Scenario #1

Questions

How would you learn about this incident? When would you learn of the incident? From whom?

What actions would you take? With whom would you need to communicate and coordinate your activities?

Does this incident have the potential to become a crisis for your organization? Why?

What policies would need to be utilized to manage this situation effectively? Do they provide sufficient guidance?

Evaluation and Action Planning Forms: Evaluation forms give players the chance to express their thoughts on important lessons as well as the

ACTION PLANNING FORM

New Company, Inc. Crisis and Emergency Management Team 90-DAY WORK PLAN		
ACTION	**ASSIGNED TO:**	**TIME FRAME:**
1.		
2.		
3.		
4.		
5.		
6.		
7.		
8.		
9.		
10.		

Figure 5-2
Action Planning Form

overall tabletop exercise process. Action planning forms are designed to help players apply the lessons they learned during the tabletop exercise to their own organization. Action Planning Forms are also essential to capture players' ideas of how to improve capabilities based on the discussions of the tabletop exercise. Action Planning form samples are provided in Figures 5-2, 5-3.

Facilitator Notes: Facilitator Notes are expressly developed to aid the facilitator in leading the discussion. The notes contain important background and

ACTION PLANNING

New Company, Inc.
Crisis and Emergency Management Team

Instructions: *During the action planning portion of today's exercise, you should focus on New Company's next steps to enhance emergency management capabilities. During this session, you will discuss the answers to the following questions and develop a 90-day workplan for taking those next steps.*

List the policies, plans, and procedures that you think should be further reviewed, supplemented, or developed. Which are priorities?

What management, support, and response capabilities are needed or should be implemented to ensure effective management of, support, and response to an emergency or crisis?

SAMPLE

Describe the action steps that should be taken in your area of responsibility. Who within your area should be assigned responsibility for each step?

What additional steps can New Company, Inc. take to enhance and maintain emergency management, support, and response capability?

Figure 5-3
Action Planning

other information discovered during scenario research and list the issues that need to be addressed by the group. These notes also indicate any unique or special aspects of the scenario, or the expected responses, that the facilitator must understand in order to properly facilitate the discussions. Facilitator Notes are best located in the front of facilitator notebooks.

Component 3 - Develop training or briefing session

A typical players' briefing for a tabletop exercise, is conducted at the beginning of the session and lasts 15–30 minutes. A briefing or training for the facilitator and documenter/evaluator is usually not required since both are heavily involved in the design, development, and preparation. However, each person should be selected with established skills and expertise for their role.

Component 4 - Develop and arrange logistics

The logistic requirements for a tabletop exercise are minimal and may only include setting up a single conference-type room with limited equipment and refreshment requirements. The room should comfortably support the number of players and observers and the facilitator/evaluator team. Equipment needs usually include only a flip chart or white board and markers.

Component 5 - Develop the final product

Proper packaging of the tabletop exercise is important. After the scenario has been finalized, personnel can begin producing the material in the format established during the design phase. The final result is packaged tabletop exercise materials ready for distribution to the players.

Part of the value of the tabletop exercise format is to solicit and evaluate spontaneous responses by the players. For that reason, the scenario(s) should remain unknown until each is presented. This can be accomplished by sealing each scenario or move, or by handing them out in sequence.

How To Develop A *Drill Or Exercise*

Component 1 - Develop the scenario

Development of a drill or exercise scenario follows the same steps as those of a tabletop exercise, but the process is more complex. The elements of a drill or exercise scenario include a narrative, timeline, messages, and message summary. The following describes the five steps of scenario development for a drill or exercise. The scenario development team for a drill or exercise is responsible to complete the scenario development and should involve focused, specifically selected personnel.

Research: As with a tabletop exercise, research should be conducted to gather technical and external information. Research is important throughout the development process, from supporting the initial selection of a scenario to providing technical information, e.g., how substances will disperse or react, what legislative initiatives would cause the incident to have high visibility with politicians and the public.

The extent of research for a drill or exercise can be quite extensive and should thoroughly explore the varied stakeholders' reactions and perspectives likely to result if the event were real. As a result, a greater number of trusted agents is usually required to support a drill or exercise than is required for a tabletop exercise, although the amount of involvement for each may be limited.

Drafting and Integration: Scenario drafting usually occurs by developing an outline and eventually a timeline of major events. The outline will eventually become the narrative. The technical details and research information will be added to the timeline and eventually used to generate specific messages. Each message is designed to replicate actual types and volume of contact with responders as well as to ensure that the major events of the scenario are properly driven. Appropriate player actions should be anticipated and planned for in the scenario materials. Finally, the message summary simply provides a chronological listing of all messages developed. All materials are developed with the support of the scenario development team whose members will develop materials individually and, as a group, integrate the total time-sequenced scenario.

Review: It is important to schedule a significant amount of time to allow trusted agents and selected simulators to review the narrative, timeline, and especially the messages to ensure that all technical and external aspects have been addressed. Several reviews will likely be necessary, and selected personnel (typically the team leader and the coordinator) will need to review the entire scenario package for overall flow and integration.

Talk-through: Conducting a talk-through is usually standard practice for a drill or exercise. For a typical drill or exercise, the talk-through occurs in the form of a working meeting. It encompasses a comparison of the messages with the narrative and timeline of events. Trusted agents and simulators work through the scenario as a team. It is the final refinement of the scenario materials and allows timing, integration, and accuracy issues to be identified prior to conduct.

Finalize: Finalizing the drill or exercise scenario can be a time-consuming process due to the volume of materials. Although the process simply involves making final edits, incorporating comments, and developing final materials, ample time should be allotted in your work plan.

A sample exercise narrative summary, timeline, message summary, and messages follow.

Sample

Narrative Summary

Date:
Thursday, November 6, 1995

Time:
At the junction of the Texas City and Houston Ship Channels, one-half mile off the northeast shore of Pelican Island.

Initiating Event:

The *Barren Sea* (a third-party tanker under contract to Oil Company X) is loaded with 460,000 barrels of Nigerian crude and en route to the company's refinery in the Texas City area. Jordan Marine International's *Jordan Woodside*, departing from Gallena Park with a crew of 25, is loaded with straight run gasoline on a charter for Starlight Oil Company. The two ships are approaching each other at the junction of the Texas City and Houston Ship Channels. The *Barren Sea* has a steering casualty and veers hard to one side into the path of the *Woodside*. The *Woodside* is unable to avoid a collision and slams into the side of the *Barren Sea*.

Unfolding Situation:

Upon collision, the two ships become locked together — the bow of the *Woodside* is locked into the side of the *Barren Sea*. The *Woodside*'s anchor drops on impact. The force and heat from the collision causes an explosion in the cargo area of the *Woodside*, which ruptures one of its cargo tanks. Straight run gasoline spills from the tank. It is ignited by sparks emanating from the grinding of metal between the two attached ships. The ships twist in the current and eventually separate.

Two of the *Barren Sea*'s oil tanks are penetrated from the impact and gashed/split vertically at the water line. Crude oil immediately gushes out of the ruptured tanks, spills into the bay water, and rises to the water surface. Crude oil quickly spreads around the damaged side of the *Barren Sea*, creating an oil slick that begins flowing towards the tidewater.

Straight run gasoline from the *Woodside* ruptured cargo tank also spills into the bay water, forming a slick that merges with the oil from the damaged *Barren Sea*. The fire on the *Woodside* spreads quickly, igniting the spilled oil and engulfing both ships in flames. Despite the fire, large quantities of crude escape the immediate area of the ships. As the fire burns, oil is washed away from the ruptured tanks, creating an elongated slick emanating from the leaking, flaming *Barren Sea*. Fire from the burning ships spreads outward to about 200 yards from each ship, but the oil slick continues out further. Over the next 45 minutes, explosions occur on board the *Woodside*.

Upon initial explosion on board the *Woodside*, the Coast Guard Duty Officer of the Port of Galveston feels the office building shake and hears a sound like low thunder. He is shortly informed of sightings of a fire in the Bolivar Roads Channel. Within 15 minutes, all movement in and out of the Houston and Texas City Ship Channels is suspended. The Captain of the Port of Galveston activates the Coast Guard's Gulf Strike Team, which arrives approximately eight hours later. In addition, the Captain of the Port (COTP) calls upon Oil Company X to mobilize all available resources to respond immediately to the incident and seeks assistance from Starlight Oil. Additional harbor tugs and salvage tugs are called to fight the fire, some of which arrive on the scene three hours later.

The *Barren Sea* drifts in the direction of the tide and runs aground in the mud. The *Woodside* continues to burn for 36 hours. As firefighters are able to get close enough to the ships to pump water on them, the engine room of the *Woodside* floods, causing it to sink stern down, further blocking the channel. Nothing can move in or out of the Houston and Texas City Ship Channels.

Three men on the *Woodside* are killed during the collision and six men jump overboard following the fire and explosion. Ten men aboard the *Barren Sea* are incinerated when their lifeboat is overrun by burning fuel. Shortly after the initial explosions, both ships are abandoned.

An oil slick spreads over the bay in the direction of the tidewater flow, presenting environmental threats.

Commercial fishing, such as shrimp harvesting, is suspended by the federal Fish and Wildlife Service. Wildlife, such as the rare sea turtles on Pelican Island, are threatened. Beaches along the bay shore are faced with oil contamination.

A national news story "exposes" the negative effects that the closed-down Ports of Houston and Texas City will have on the availability of winter fuel. Consumers rush to the gas stations, filling their vehicle gas tanks and spares, and arrange heating oil fill-ups to avoid the perceived supply shortages and price jumps.

Government regulators take a hard line against Oil Company X to firmly establish the new Oil Pollution Act (passed on August 18, 1990). There is pressure from Washington to make an example out of the company responsible for the first major oil spill since the new legislation.

(This is a hypothetical situation for exercise purposes only.)

Sample

Incident 2 Scenario Timeline

Thursday, November 6

3:30 a.m. Oil Company X's *Barren Sea* and the *Jordan Woodside* collide and lock together in the Texas City and Houston Ship Channels. *Woodside*'s anchor drops. Three men on *Woodside*'s bow are killed on impact.

3:35 a.m. *Woodside* has an explosion on-board. Tank ruptures and gasoline spills into the bay. Fire spreads to floating gasoline. Six men on *Woodside* jump overboard.

Coast Guard and Oil Company X in Texas City are alerted through distress calls from other vessels.

3:37 a.m. *Barren Sea* spills major amount of crude from two tanks.

3:42 a.m. Tug, *Wontsink*, pulls along side of *Barren Sea* to assist.

Burning gasoline ignites crude oil and spreads to *Barren Sea.*

3:47 a.m. Coast Guard helicopter, returning from a search and rescue mission, flies over the scene. Large fire seen but little else.

3:50 a.m. Coast Guard notifies Oil Company X's Transport Duty Officer in Denver of the collision.

Wontsink casts off due to hazardous situation.

Oil Company X Transport begins internal notifications.

3:55 a.m. Coast Guard vessel arrives on scene and rescues crew members off burning ships.

4:00 a.m. (COTP), Galveston, from home, closes all surrounding channels and waterways.

Oil Company X's command post staff calls out salvage/firefighting and oil pollution contractors and puts clean-up contractors on notice.

Oil Company X's command post staff requests Coast Guard response team assistance.

Oil Company X calls out salvage specialists and clean-up contractors. Houston Vessel Traffic System transmits traffic advice to affected ships.

Coast Guard makes governmental notifications and requests Coast Guard and contractor firefighting/clean-up assistance.

4:10 a.m. Fire is out of control. All hands abandon ship. Lifeboats on both ships are launched.

4:15 a.m. Engine of one *Barren Sea* lifeboat fails. Boat and crew are overrun by burning oil and incinerated.

COTP arrives at Galveston office from home.

Oil Company X's On Scene Commander (OSC) calls out command post staff and plans to establish a post at the refinery in Texas City.

4:30 a.m. The ships twist in the current and separate. The *Barren Sea* begins moving in the direction of [weather dependent].

4:45 a.m. Oil Company X's command post staff begins arriving at their post.

Company X-contracted firefighting tugs begin arriving on scene.

5:00 a.m. *Barren Sea* runs aground at [weather dependent].

... continued

(This is a hypothetical situation for exercise purposes only.)

Sample

Message Summary

Thursday a.m.

MSG#	TIME	FROM	SUBJECT
X-OIL-1	3:30 a.m.	XPL, Denver	Initial notification
X-PR-1	3:40 a.m.	PR, Texas City	Notification of spill
X-NEWS-1	3:45 a.m.	PR, Texas City	Inquiries
LAW 1	3:49 a.m.	XPL, Denver	Notification
X-EH&S-1	3:49 a.m.	XPL, Denver	Notification
X-OIL-2	3:49 a.m.	XPL, Denver	Notification
X-PR-2	3:49 a.m.	PR, Texas City	Evacuation of affected area
X-EH&S-2	4:02 a.m.	XPL, Denver	Location of break
LAW 2	4:02 a.m.	XPL, Denver	Incident update
X-OIL-3	4:10 a.m.	XPL, Denver	Incident update
X-PR-3	4:20 a.m.	XPL, Denver	Incident update
X-PR-4	4:25 a.m.	PR, Texas City	Mayors' request for briefing
X-OIL-4	4:35 a.m.	Merrillville	Leaving for site
X-PR-5	4:55 a.m.	PR, Texas City/Site	TV arriving
X-PR-6	5:00 a.m.	PR, Atlanta	Media calls
X-OIL-5	5:05 a.m.	Atlanta	Incident update
X-PR-7	5:05 a.m.	PR, Texas City/Site	Level 3 emergency

Sample

Exercise Message

__X__PHONE   MSG# X-OIL-1

______FAX   DATE: Wednesday

______COMPUTER   TIME: 3:30 a.m.

______IN PERSON

______RADIO/OTHER

FOR: Company X Notification Center

FROM: Area Manager, XPL Corporate, Denver

MESSAGE: **THIS IS AN EXERCISE**

We have just received notification from the Galveston COTP that our ship, the *Barren Sea*, has been involved in a collision with another ship, the *Jordan Woodside*, in the Houston Ship Channel. He reports that there has apparently been some sort of explosion and a fire. However, at this time the extent of damage is unknown. We have been trying to reach the captain of the ship by radio but so far have been unsuccessful. We are maintaining contact with the COTP who is in the process of dispatching personnel to the scene. We will also continue to try and make radio contact with the captain of the *Barren Sea*. As soon as we receive any update or further word on the situation, we will call you.

THIS IS AN EXERCISE

CONTROLLER/SIMULATOR INSTRUCTION: Note attached incident response form.

DOCUMENT ALL ACTIONS RELATED TO THIS MESSAGE ON REVERSE SIDE

Sample

Exercise Message

__X__PHONE	MSG# X-PR-1
______FAX	DATE: Thursday
______COMPUTER	TIME: 3:40 a.m.
______IN PERSON	
______RADIO/OTHER	

FOR: Dan Frampton, PR Corporate, Denver -AT HOME

FROM: Jane Smith, PR, Texas City (from home)

MESSAGE: **THIS IS AN EXERCISE**

Dan? I'm sorry to have to wake you, but one of our ships, the *Barren Sea*, has collided in the Houston Ship Channel with another ship, the *Jordan Woodside*, I believe, and it's a bad one. I'm sure we've got some sort of a spill on our hands, but I don't have all the details. I'm sure that we can expect a million phone calls and I wanted to give you a heads-up. I'm getting ready to head to the office.

I'm thinking of calling Miles Bliven, since he's closer than Dave and should be able to handle this quite well. I don't want to throw [the new PR rep in Denver] into this, since he/she just started his/her job there. Who else do you think I should notify? How often do you want me to call you with a report?

THIS IS AN EXERCISE

CONTROLLER/SIMULATOR INSTRUCTION: Simulator should tailor future messages to comply with this request to the last question.

DOCUMENT ALL ACTIONS RELATED TO THIS MESSAGE ON REVERSE SIDE

Component 2 - Develop support materials

For ease of use, drill or exercise support materials are usually contained in a handbook for each controller, evaluator, and simulator. The different sections of the handbook are described here.

Schedule of Activities Section: As discussed in the design phase, the schedule of activities lists all pre-conduct, conduct, and post-conduct activities for the conduct organization. Sample schedules are contained and discussed in more detail in Chapter 6.

General Information Section: General information includes, as background information, overviews of the specific exercise activity as well as an overview of the organization's overall crisis and/or emergency management program. This chapter also includes a listing of the response locations that are expected to see play during the drill/exercise ... and a list of points of contact should there be any questions regarding the drill/exercise.

Conduct Organization Section: In the conduct organization section, you will find a description of each conduct position (discussed in detail in Chapter 7). Also included is the drill/exercise organization chart(s) and controller/evaluator assignment locations. The controller/evaluator assignment locations list identifies each controller's, evaluator's, and simulator's assignment location, when they should be there, and identifies their position. A sample exercise organization chart is contained in Figure 5-4. A sample assignment list form is depicted in Figure 5-5 and a sample assignment list follows below.

Sample

Oak Hill Limited Exercise

Dayton Organization:

Name	Position	Location	Phone #
Troy Ketchum #1	Exercise Coordinator	all access	555 412-8820
Mike Mather #2	Exercise Coordinator	all access	555 412-8820
William Stevens #1	CMT E	CMT, 30th floor	contact via B. Jacobs
Bill Jacobs #1	CMT C/E	CMT, 30th floor	555 412-8821
Marty Williams #1	P&GA C/E	P&GA, 37th floor	555 412-8822
Trey Wilson #1	HR C/E	HR, 36th and 38th floors	555 412-8823
Mark Winston #2	HR C/E	HR, 36th and 38th floors	555 412-8823

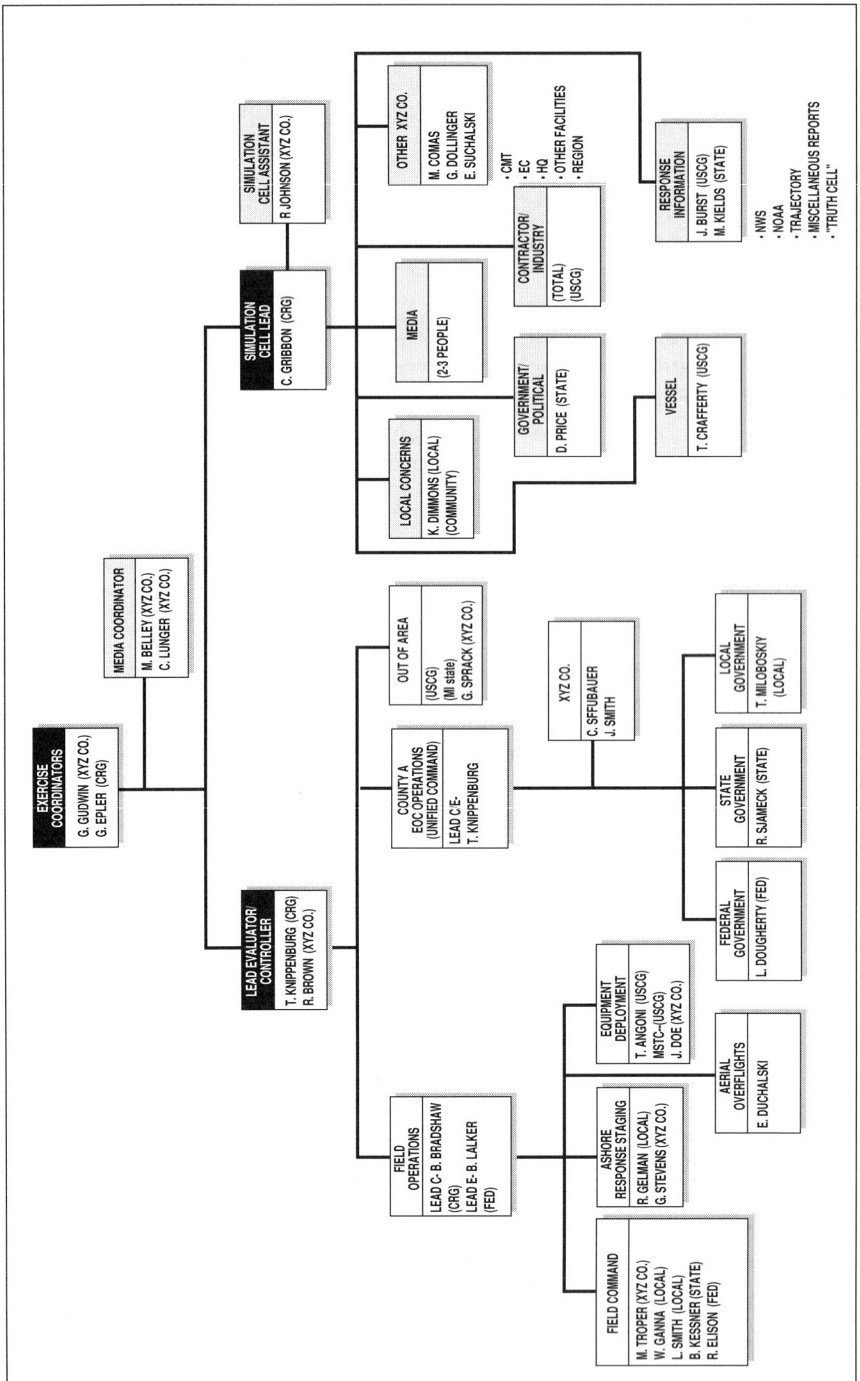

Figure 5-4
Company XYZ Prep Area Exercise Conduct Organization

CONTROLLER/EVALUATOR ASSIGNMENT LOCATIONS

Use the following form to develop your assignment list and locations

Name and Geographic Location (e.g., Dallas, UK)	**Response Site Location** (e.g., EOC, or Command Post)	**Assignment** (e.g., C/E/S/O)	**Conduct Phone Number or Radio Frequency**

Figure 5-5
Controller/Evaluator Assignment Locations

Name	Position	Location	Phone #
Steve Daniels #2	floater C/E	EH&S, 48th and 49th, Law, 20th and 21st floors	555 412-8822
Tim Wilks #1	Law C/E	Law, 20th and 21st floors	555 412-8824

#1 = Shift (1pm–12am) and (10am–4pm)
#2 = Shift (11pm–11am)

C = Controller E = Evaluator
S = Simulator

Important Telephone Numbers: A telephone directory is also normally located in the handbook. The telephone directory is critical for simulators and controllers to function effectively. The controller/evaluator/simulator telephone directory includes a list of role-players with their designated roles and assigned numbers as well as critical organization numbers such as the lead controller's contact number and lead evaluator and simulator contact numbers. The following is a sample telephone directory.

Sample

Player Positions and Locations

1. Fairfield

Department	1st Shift	2nd Shift	Ext.
EXECUTIVE GROUP:			
Executive Group Leader (EGL)	Robin Swan Tony Devine	Tim Welsh	4500
Office of General Counsel (OGC)	Kay Allen	Dennyn Rakin	4500
Public Affairs (PA)	Coleman Cox	Sarah Dexter/ Coleman Cox	4500
Employee Relations (ER)	Dan Lucas Ed Hancomb	Rick Starr/ Don Beyer	4500
Treasurers	Gary Roth	Jim Cassell	4500
Government Relations	Sue Glennis	Steven Scott	4500
Marketing	Bob Bryan Mike Murphy	Dick Gross	4500
Medical	Mark Cannell/ Walter Conrad		4500
FAX			3608

STAFF GROUP:

Staff Group Leader (SGL)	Doug Baxter	Don Voelte	2912
Deputy SGL	Bill Hollis	Matt Anchor	2912
Operations (Ops)	Jim Conners/	Tim French/	2905
	Dana Dixon	Sam Atler	2903

Telephone directories for players are also critical. They list all simulated roles with their associated simulation cell telephone numbers. These directories are not usually included in the handbook and should be created separately and distributed to the potential players. The following is a sample simulator telephone directory.

SAMPLE

Simulator Telephone Numbers

A number of Great Bay and external audiences are being simulated for the purposes of this drill. The following telephone numbers should be used by players to reach nonparticipating persons and organizations simulated during the drill. Technically knowledgeable simulators will be role-playing nonparticipating individuals, organizations, and locations.

If you have any questions regarding simulation or which telephone numbers to use, contact the nearest controller/evaluator or call the main simulation cell number. These numbers will only be active during drill conduct.

Great Bay Personnel/Organizations	
Affected Great Bay site operation:	517-1893
Emergency Response Strike Team:	517-1893
Other Great Bay operations within Portsmouth:	976-5028
Other Great Bay operations outside Portsmouth:	976-5028
Concerned Great Bay employees:	976-5025
External Groups	
Contractors/other industry members:	976-5028
Local government officials:	976-5050
State government officials:	976-5027
Federal government officials:	976-5027
Media:	976-5025/5027
Trade associations:	976-5025
General public:	976-5025/5050
Other Parties	976-6097

Guidelines Section: Guidelines describe the expectations for those supporting conduct. These guidelines will be discussed in more detail in Chapter 7. They cover guidelines for controllers, evaluators, and simulators.

Real Events Section: This is an extremely important section of the handbook. It lists the name(s) and number(s) of personnel to call should any real emergencies or real media calls occur during conduct of the drill or exercise. It should state the process to notify the proper personnel of a real situation. It should also describe the decision process with regard to its impact on the drill or exercise conduct.

Objectives & Scope Section: The objectives and scope section encompasses information developed during the design phase and contained in the design document, including the overall objectives of the drill or exercise, the parameters, and the extent of play. This information is in the handbook to provide the context and limits of the drill or exercise for the controllers, evaluators, and simulators.

Scenario Information: The scenario information includes all the materials you developed in the previous step of development. These items include the scenario narrative, timeline, messages, and message summary. When packaging the scenario information, it will help conduct personnel if you include in their books only the materials that are pertinent to their roles. This requires additional effort to package the handbooks but makes the job easier for conduct personnel.

Records and Forms: The typical records and forms included in this section are evaluation forms with instructions, observation log forms, blank message forms (for simulators only), call log forms (for simulators only), player evaluation forms (for controllers only), and debriefing guidelines (for controllers only). These forms will be discussed in more detail later in the book.

Component 3 - Develop training or briefing sessions

When conducting a briefing for a drill or exercise, here are some issues you should consider.

- Is the number of personnel large enough and their schedules diverse enough to require multiple sessions?
- Should briefings be conducted in multiple locations?
- What type of support materials will be needed?
- What are the logistic requirements (i.e., rooms, projectors, etc.)?

You should also decide if observers will be present and whether separate observer briefings will be held. Some advantages of including observers are

- they add another set of insights to the evaluation,
- they can spread the lessons learned throughout the organization,
- their participation can be a sign of goodwill.

Some disadvantages of including observers are

- they can interfere in conduct if not managed correctly or given proper guidance,
- they may distract or take away from the realism of a drill or exercise (i.e., too many people).

Component 4 - Develop and arrange logistics

In a drill/exercise, the logistical requirements are much more complex. Some logistical items you will need to consider when planning a drill or exercise are

- training and briefing needs (including facilities, equipment, and training/briefing support materials);
- security or access arrangements for appropriate personnel;
- simulation cell needs, which will be discussed in detail in conduct;
- types of communication equipment needed during the drill or exercise;
- awareness policy, to include signs distributed throughout the playing areas to indicate that a drill or exercise is underway;
- possible refreshments and/or meals for the conduct organization.

Component 5 - Develop final product

Due to the large amount of scenario materials created for a drill or exercise, it is important to identify your production process during the early stages of design and development. As mentioned previously, a binder is frequently the best means of packaging the materials for a drill or exercise. The finalization process involves duplicating all relevant materials, including the scenario materials, into the final packaging. Consideration should be given to whether everyone in the conduct organization will receive all material or whether different groups or individuals should receive only selected materials. These decisions may dramatically impact your production time and requirements. Your process should also include a quality assurance process. Because of the volume of materials that will need to be produced and checked for quality, it is important to allow sufficient time in your workplan for this task.

HOW TO

Develop Your Exercise Activity

☐ *Select, Research, and Develop the Scenario*

- *Research*
- *Draft and integrate*
- *Review*
- *Talk-through*
- *Finalize*

☐ *Identify Support Materials Required*

- *Player groundrules/guidelines*
- *Forms*
- *Player Tools*
- *Evaluation Tools*

☐ *Define and Develop Training and/or Briefing*

- *Identify audiences/groups*
- *Assess level of experience*
- *Outline role/training requirement*
- *Define briefing information*
- *Develop training/briefing materials*
- *Schedule session(s) and arrange logistics*

☐ *Identify and Arrange Logistics to Fully Support Conduct and Evaluation*

- *Arrange travel, facilities, equipment, and scheduling*
- *Obtain resources, reproduction support, etc.*

CHAPTER 6

Creating Reality With Conduct

Effective Crisis Management is often a function of effective teamwork. The bonding of a team is enhanced when its members exercise together. This allows everyone to begin to understand how they will work together under the pressure of "the real thing'.

Mayer Nudell
Author
Handbook for Effective Emergency and Crisis Management

This chapter reinforces the necessary groundwork for effective conduct of an exercise activity and details how conduct is carried out.

The objectives are to help you

- understand all the elements involved when conducting any exercise activity,
- understand how exercise activities are conducted,
- be able to construct a typical exercise activity conduct schedule.

Introduction

The actual conduct phase of an exercise activity consists of initiating play, simulating and controlling or facilitating activities, and monitoring to ensure that play remains within the design parameters and that appropriate issues are raised. It also involves documentation and initial evaluation of player response and termination of play, as appropriate.

Purpose of Exercise Conduct

The conduct phase presents the culminating activity for players. The primary conduct phase components are depicted in Figure 6-1. In addition to the steps initiated by the conduct organization, conduct provides an opportunity for player training and for validating knowledge and/or skills regarding the organization's crisis and emergency capabilities, plans, and procedures. Due to the very distinct nature of tabletop

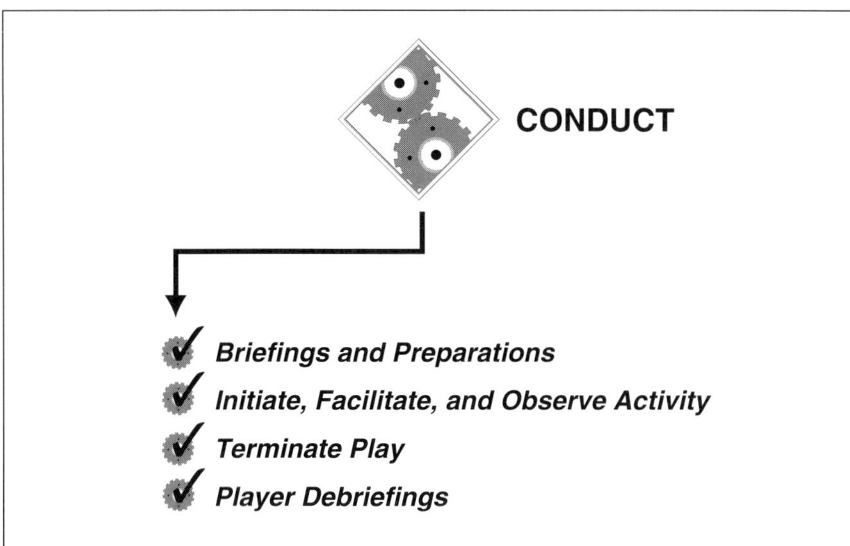

Figure 6-1
The Conduct Phase

exercises versus drills and exercises, this chapter discusses them separately, focusing on the unique aspects of conduct. In addition, the format answers the what, why, and how of each step.

Conduct of a *Tabletop Exercise*

Tabletop exercises are short simulation activities that allow discussion of either a single or several hypothetical scenarios. Conduct of these sessions tends to be relatively standardized and can follow a common agenda. Tabletop exercises usually last from four to eight hours, including briefings, facilitated discussion, and action planning sessions.

This timing is based on

- the typical scope and objectives for the tabletop exercise,
- the type of format, (i.e., where there is discussion, not action),
- the optimum concentration span of a group,
- time availability of participants.

The following are sample agendas for the two types of tabletop exercises (multiple scenario and single scenario with moves). The first is an agenda for a session utilizing multiple scenarios.

Tabletop Exercise Agenda

1:30 p.m.	Introduction and Ground Rules
1:45 p.m.	Scenario One, Facilitated Discussion
3:00 p.m.	Scenario One, Evaluation and Discussion
3:30 p.m.	Scenario Two, Facilitated Discussion
4:45 p.m.	Scenario Two, Evaluation and Discussion
5:15 p.m.	Overall Evaluation and Findings
5:45 p.m.	Action Planning Session

The following is a typical agenda for a tabletop exercise that involves multiple moves in a single hypothetical scenario.

Tabletop Exercise Agenda

8:00–8:15 a.m.	Introduction and Ground Rules
8:15–9:45 a.m.	Scenario Background
9:45–10:45 a.m.	Move One
10:45–11:00 a.m.	Break
11:00–12:00 a.m.	Move Two
12:00–1:00 p.m.	Lunch and Action Planning Session
1:00 p.m.	Conclusion

In order to understand fully the stages of conduct, it is helpful to discuss each component of the agenda individually. The following details what is involved in each component.

Introduction and Ground Rules of Play

Time Scheduled: 15–30 minutes

What?

The **introduction** is normally utilized to

- act as a transition from another activity prior to the tabletop exercise (e.g., training),
- introduce players and facilitator(s),
- briefly describe the tabletop exercise process,
- briefly describe the flow of the exercise,
- give any special instructions for the tabletop exercise (i.e., if there is going to be more than one scenario; if you should wait until proceeding to another chapter, etc.),
- describe the purpose of activities such as the action planning session.

The **ground rules** are instructions for the players. The ground rules usually address

- the objectives of the tabletop exercise,
- the importance of actively participating in the tabletop exercise (i.e., how to get the most benefit out of the activity),
- issues that players should consider during scenario discussions.

Why?

The introduction is important because it sets the stage for facilitated discussion. It establishes players' expectations by stating what will be accomplished during the session. An introduction should put the players in a relaxed, stress-free atmosphere that allows for free flow of thought on the part of the players.

The ground rules are the rules of play and player instructions outlining what is expected from the players during exercise play. By establishing them up front, it helps players get as much out of the session as possible.

How?

The facilitator or one of the participants provides the introduction. The facilitator should present the ground rules, leaving time for any questions the players may have. The ground rules should be presented in written form as well as orally.

Scenario Presentation

Time scheduled: 15–30 minutes per scenario or move

What?

The facilitator presents the scenario background, i.e., the description of the incident and any other pertinent background information. The players then briefly digest the information and individually complete a brief question-and-answer session aided by specially-designed question forms.

A tabletop exercise may have more than one scenario or more than one move per scenario, depending on the established objectives and format in which you are working. Each scenario or move is presented separately at the time it is to be discussed.

Why?

The scenario is the essential tool used in tabletop exercises to bring out the important issues that need to be discussed.

Players fill out question forms during the presentation to help them formulate and focus their thoughts and allow a greater freedom of expression than may be possible in a general discussion. The facilitated discussion should not necessarily follow the flow of these questions.

How?

You can present a scenario

- verbally,
- in written form (narrative or timeline of events),
- in videotape format.

It is recommended that all three formats be used in combination when presenting the scenario.

You can present the entire scenario at once, or you can present a move at a time, if appropriate.

- A single scenario with one or more moves is optimal when discussing phases of operations or when different response actions occur during a running time sequence. This method is useful as a training tool when preparing for a larger scale drill or exercise.
- Multiple scenarios are useful to introduce players to a wide range of incidents for which they can contrast and compare response activities. This method is ideal for an organization in its early stages of crisis or emergency management planning and response. Senior level officials also find it useful when developing and revising existing response plans and procedures.

Facilitation and Discussion of Scenario(s) and Move(s)

Time Scheduled: 1 hour to 1 hour and 45 minutes per scenario or move

What?

The facilitator leads the players through a discussion of the scenario or move in order to address the established issues/objectives.

Why?

This is a crucial time during the tabletop exercise. It is during this time that the tabletop exercise objectives are addressed. This segment of the tabletop exercise can also aid in

- discovering areas for improvement within planned response,
- building consensus,
- reinforcement of roles and responsibilities,
- validating knowledge of response expectations and requirements,
- providing valuable learning opportunities,
- clarifying issues and policy.

How?

Each move or separate scenario is usually discussed individually with breaks between them.

The facilitator, using the prepared facilitator notes, should enlist thought-provoking questions to lead the group discussion and to prompt consideration of critical points. The facilitator also injects probing questions or comments into the conversation, as appropriate. The facilitator helps to create an atmosphere that allows for free exchange of thought by the players while being careful not to allow the discussion to stray too far from the original objectives.

The amount of time players are given to free play and brainstorm ideas versus the amount of directed discussion should be decided based upon the objectives and the weight given to training versus validation.

The assistant facilitator or evaluator should take notes of

- the discussion,
- decisions made,
- identified areas of uncertainty or those needing improvement,
- possible action items.

These notes and the facilitator assessment will be used during the action planning phase and as points to be reinforced at the conclusion.

Action Planning

Time Scheduled: 45 minutes to 1 hour and 30 minutes

What?

The action planning session provides a time to consolidate and prioritize what players view as the next steps for their organization to enhance its crisis and emergency management capabilities. During this session players also have the opportunity to address any action items brought out by the discussion session.

Why?

The list developed by participants during this session will be useful to identify next steps for the program overall and for individuals with crisis and emergency management responsibilities. It documents the "game plan" for improvements as well as the conclusions reached during the session.

How?

Facilitators guide participants to the Action Planning form, asking them to address the following questions based on the discussion:

- What capabilities should be developed/further validated?
- What policies or procedures should be reviewed or developed?
- What steps should players or designated personnel take to improve the company's crisis management capabilities?

Participants should also

- identify other action items,

- assign responsibility for each action item addressed during the discussion,
- set realistic timelines for completion of each action item.

Conclusion

Time Scheduled: 15–30 minutes

What?

This is the last part of the tabletop exercise. It allows the facilitator to sum up the important issues, make final points, and summarize future action items.

Why?

This reemphasizes the important issues and objectives that participants should take away from the session.

How?

The facilitator should review the objectives and how they have been met. The facilitator should also summarize the discussion highlights and the action item list created during the action planning session and conclude with his/her assessment of the session.

Conduct of a Drill or Exercise

Drills and exercises are simulation activities that allow actual demonstration of response capability. The range of complexity, level of detail, and emphasis on training versus capability validation varies greatly. Nonetheless, the activities and timing associated with conduct of these sessions tend to be relatively standardized, with the exception of the timing of play (i.e., between initiation and termination). The total time involved in the conduct phase for a drill or exercise can typically range from 4–72 hours depending upon the

- objectives and scope,
- number of groups and participants,
- exercise program and activity objectives.

A typical conduct schedule for a drill or exercise follows.

Sample

Schedule Of Activities

Activity	Date & Time	Location
Player Briefings	4/14 2:00 p.m. 4/14 2:00 p.m. 4/16 8:00 a.m.	Houston: All Players, Room 301-304 Newark:South Hall Cafeteria Galveston: Beach Building, Room 700
Simulation Cell Setup and Communications Check	4/20 a.m.	Houston:Simulation Cell, Room 213/214 Newark:Simulation Cell, North Hall, Room 80
Observer Briefing	4/20 10:30-11:30 a.m.	Houston:Simulation Cell, Room 213/214 Newark: Simulation Cell, North Hall, Room 80
Lunch and Tour	4/20 11:30-1:00 p.m.	Houston: Simulation Cell, Room 213/214 Newark: Simulation Cell, North Hall, Room 80

Player Briefings

Time scheduled: one to two hours, usually one to two weeks prior to the day of the drill or exercise but can occur the day before.

What?

A meeting to brief players on the ground rules of conduct. The briefings usually cover the overall objectives and scope of the drill or exercise, the parameters and limits of play, simulation plans, and how the debriefing process is going to occur.

Typical information covered during the briefing includes

- overview of the company's crisis and emergency management program and the specific response team(s)' concept of operations, if necessary;
- description of drill or exercise scope and parameters;
- drill or exercise objectives;
- player guidelines (i.e., "This is a Drill/Exercise," whom to call, how to communicate, debriefing process, etc.);
- the function of the simulation cell;
- what to do if a real event occurs;

- what to do if real media representatives call;
- where to get further drill or exercise information, before, during, and after conduct;
- questions and answers.

Why?

This step is strongly recommended. Even though participants may already be familiar with drills/exercises, no two drills or exercises are the same, and the player briefing is proven to improve the quality and effectiveness of your activity.

Considerations for holding player briefing(s):

- Personnel have never or infrequently participated in an exercise activity;
- The scope of the drill or exercise is bigger then personnel's experience;
- There are unique/unusual aspects to the drill or exercise, i.e., time-jumps, extensive mobilization, etc.

How?

Player debriefings are usually conducted by the coordinator, designated lead controller, or others who were involved in the design and development phases of the drill or exercise. The session is usually presented using overheads or slides with handout materials provided.

Simulation Cell Setup and Communications Check

Time scheduled: four to six hours, the day prior to drill or exercise

What?

This provides a chance for the logistics team to set up the simulation cell equipment (phones, radios, sound barriers, desks, fax machine(s), computer(s), etc.) and test all the communication equipment prior to its use.

Why?

It is important to set up the simulation cell a day prior to conduct so that simulators can become familiar with the setup during their briefings. The logistics team can also identify and resolve potential problems before they can impact the drill or exercise.

It is especially important to test the communication equipment used to establish linkages between the different groups in the conduct organization (i.e., between controller, evaluator, player, simulator). This communication flow is essential, since it is the vehicle by which the scenario is presented and controlled.

How?

During the design and development phase, you should have identified the number of simulation cells and simulators required for the exercise activity.

Personnel should set up the simulation cell(s) with the appropriate equipment, such as

- telephones for each simulator, simulation cell lead, and simulation cell assistant, if applicable;
- a desk for each simulator and simulation cell lead with paper and pens for notes;
- sound barriers for the desks;
- an overhead projector and screen;
- a fax machine;
- a flip chart, paper pad, tape, and black markers;
- relevant maps and crisis and emergency management plans and procedures (to be determined);
- name tents–place cards giving role and extension, to be placed on the desk of each simulator, simulation cell lead, and simulation cell assistant
- shredder;
- access to a copier machine.

Personnel should conduct a communications check including

- each simulator's incoming and outgoing telephone line;
- the rollover of simulators' phones to a master phone;
- controller/evaluator communication system, whether hand-held radios or cellular telephones;
- communications both inside and outside and among all possible play locations;
- the receipt and transmittal functions of the fax machine;
- the computer(s).

Controller/Evaluator/Simulator Training and/or Briefings

Time Scheduled: two to four hours, the day prior to the drill or exercise, or if conducting a night drill or exercise, the morning prior to drill or exercise

What?

These briefings are intended to be a training and information session for the controllers, evaluators, and simulators. Ideally, each controller, evaluator, and simulator should have some experience or training in their roles prior to attending the briefing. They should also be familiar with the response organization and plan. Controllers, evaluators, and simulators are briefed on their roles and responsibilities and given general drill or exercise information. The briefings also give the controllers, evaluators, and simulators the opportunity to ask any questions they may have prior to the exercise.

Why?

Briefings are conducted so that controllers, evaluators, and simulators

- know what they are supposed to do, how they are to do it, and when they are to do it;
- know how the exercise is being controlled and evaluated;
- can identify questions and resolve any uncertainty with the scenario and expected response actions before conduct;
- can be sure to understand any special requirements.

These briefings are critical to the overall exercise conduct and control. This is the opportunity to review how controllers, evaluators, and simulators have separate and specific roles and responsibilities; how their movements (especially those of the evaluators) need to be controlled in order to get an accurate assessment of the player response; how simulators add realism to the scenario, etc.

How?

You can conduct one briefing for all controllers, evaluators, and simulators together or brief each group separately. Controllers and evaluators are typically briefed together by the lead controller, while the simulators are briefed separately by the simulation cell lead.

The briefers should discuss the

- drill or exercise conduct organization;
- schedule;
- overview of scenario, scope, and objectives;
- required documentation during the evaluation process;
- overview of roles and responsibilities;
- process for problem resolution;
- communication system being used for the conduct organization;
- unique aspects of the drill or exercise;
- debriefing process.

Training and briefings should be conducted in a location that will not tip off players if the activity is unannounced.

Simulator Dry Run

Time Scheduled: one hour to one hour and 30 minutes following the simulator briefing, on the day prior to the drill or exercise. If conducting a night drill or exercise, the dry run can be held on the morning prior to the drill or exercise. Simulator dry run can also be done immediately before conduct, within one hour, but that is less desirable because it leaves too little time to resolve any problems that may arise.

What?

A dry run is a practice session where simulators make and receive calls as if the drill or exercise were actually underway.

Why?

A dry run allows simulators to become familiar with the phone and other communication systems they will be using during the drill or exercise. It also provides the opportunity to final test the communications between the simulation cell lead and the controllers as well as between the different simulation cells, if applicable.

How?

The dry run is conducted in the simulation cell and is led by the simulation cell lead, who

- directs the dry run,
- observes calls placed to/from each simulator,
- provides guidance and assistance to simulators.

Simulators can use messages created especially for the dry run or existing drill or exercise messages. Simulators can make calls to each other or, preferably, the controller/evaluator organization can assist in the dry run by making calls to, and receiving calls from, the simulators. Role playing during these calls is just as important as during the drill or exercise.

Observer Briefings

Time scheduled: one-half to one hour immediately prior to the drill or exercise

What?

A meeting to familiarize observers with the design and scenario of the drill or exercise as well as to cover the guidelines for observing.

Why?

If you have observers, the briefing is important because it

- gives the observers an idea of what they will be seeing;

- explains the ground rules to them, limiting the chance of observer interference during conduct.

How?

An observer briefing is usually conducted by the coordinator, a lead controller, lead evaluator or, preferably, a designated observer manager. The briefing should cover

- the objectives, scope, and scenario of the exercise activity;
- the observer ground rules;
- the schedule for the observers (if applicable);
- how they can give input into the evaluation process.

Tours of Area of Play

Time scheduled: one hour to three hours

What?

You may want to give controllers, evaluators, simulators, and observers a tour of the area in which play may occur or where the majority of response action will take place. Tours are usually conducted in conjunction with briefings but should be coordinated so that players are not tipped off if the activity is unannounced.

Why?

Though this is not a required step, it does allow the drill or exercise conduct organization to become familiar with the facility that they must evaluate, control, or drive through simulation. It also provides a context and mental picture of where response will take place.

How?

A tour should be given by personnel familiar with the facility, the scenario, and the scope of the exercise activity. This should not be conducted by someone who is expected to play as a participant in the drill or exercise.

Report for Conduct

Time Scheduled: 30 minutes prior to drill or exercise initiation.

What?

Controllers, evaluators, and simulators report to the simulation cell for last-minute guidance and questions before the drill or exercise begins.

Why?

This gives the coordinator and other leads the opportunity to reinforce or add information in addition to giving the controllers, evaluators, and simulators the opportunity to ask any last-minute questions. It also gives the drill or exercise organization time to settle into their positions or prepare to be dispatched before the drill or exercise starts.

How?

The coordinator and simulation cell lead and/or each designated lead should gather his/her designated group of people together prior to the drill or exercise to settle any outstanding issues. Additional materials are also provided at this time, and from here controllers and evaluators are dispatched to the appropriate response location just prior to initiation of play. In order to access and control the total response operationaccurately, the controllers and evaluators need to be at their respective locations just prior to the players' arrival.

Drill/Exercise Play

What?

This is when the actual conduct of the drill or exercise takes place, from initiation to termination of play.

Why?

Drill or exercise play allows players to respond and manage the unfolding situation, meet the objectives of the drill or exercise, demonstrate actual capabilities, and experience the validation and training process.

How?

Play is initiated and driven by simulators (and occasionally controllers) who

- input pre-scripted messages (via telephone, in person, computer, fax, radio, and other, as directed) at a designated time so that the scenario progresses in a realistic manner,
- respond to requests for information from players within designed/realistic response parameters,
- adjust the scenario as directed by injecting new information or questions or modifying existing information.

Play is controlled/directed by

- controllers—who observe play and give feedback to the simulators, who can then modify, redirect, or adjust the pace of the scenario;
- controllers—who intervene with play only if absolutely necessary to keep the activity within the defined scope;
- simulators—who track the progress of the scenario using status boards and a posted timeline of events.

Drill or Exercise Termination

What?

Termination is the decision and notification of drill or exercise personnel and players of the end of the drill or exercise.

Why?

It is important to announce the termination of play at all sites at one time, to ensure that play does not continue beyond a necessary point and to ensure that play does not stop before it is appropriate, in order to allow all objectives to be met.

How?

A general termination time is usually decided upon during the development phase. This termination time can be revised during conduct if it is thought that the objectives have not yet been met, or if they have been met sooner than expected and play is no longer required.

The final termination time is determined by the coordinator and is announced simultaneously by the controllers and simulation cell lead(s) at the designated time.

Player/Controller/Evaluator/ Simulator Debriefings

Time Scheduled: one hour to one hour and 30 minutes

What?

Debriefings are meetings held after the drill or exercise to capture the results from the players and conduct organization.

Why?

The controller, evaluator, and simulator debriefings help ensure that a thorough and coordinated evaluation is possible. They also allow integration to be discussed and open observation issues to be resolved.

Player debriefings allow players to give feedback on their performance during the drill or exercise, identifying lessons learned and possible improvements. Feedback from the players is also solicited in order to improve the exercise process.

How?

Debriefings are usually conducted in each location as soon as possible after the drill or exercise, either immediately following termination or the following day.

Player debriefings are conducted by the lead controller at each player location. The controller/evaluator debriefings are conducted by the lead controller, and the simulator debriefing is conducted by the simulation cell lead. The person conducting the debriefing can also be responsible for documenting the session, or another person can be assigned that responsibility. Evaluation and critique forms are also completed and returned prior to the departure of the players.

Topics covered during the debriefings are

- observations of positive actions taken by players and possible improvements;
- performance against objectives and defined evaluation criteria, response plans, and standards;
- issues or response activities that should have been addressed during the drill or exercise, but were not;
- comments on the drill or exercise process.

Simulation Cell Breakdown/ Packing of Materials

Time Scheduled: The day after conduct so that the organization has enough time to gather all the necessary documentation for the evaluation.

What?

The gathering up of all relevant forms, destroying all unnecessary materials and breaking down the equipment to return the room(s) the original condition, and to return equipment and supplies.

Why?

To return to normal operations as quickly as possible and to get the room and equipment back into service to support day-to-day operations.

How?

When gathering and packing materials and breaking down operations, you should

- gather all evaluation forms and player materials and label correctly;
- destroy all unnecessary paper (possibly by shredder);
- disconnect and return all equipment to its original place;
- pack up any extra administrative supplies and return to its rightful place;
- ensure that a cleaning crew has been scheduled, if necessary;
- return any maps or plans to the correct department/person.

Conduct is the main event of the drill or exercise process, but it is not the final step. Ensuring a complete and thorough evaluation is the next important step in the exercising process.

HOW TO

Plan for Conduct

- ☐ *Consider (visualize) how conduct will be accomplished*
- ☐ *Ensure an appropriate schedule is prepared, outlining all conduct activities*
  - *- Briefing/training sessions*
  - *- Arrival time(s) for conduct organization*
  - *- Duration of exercise activity*
  - *- Action planning session(s) and/or debriefing sessions*
- ☐ *Ensure all members of the conduct organization are carefully selected*
- ☐ *Prepare facilities and other setup requirements*
- ☐ *Initiate play*
- ☐ *Coordinate and announce termination*
- ☐ *Conduct debriefing sessions*
- ☐ *Carry out the conduct*

CHAPTER 7

CONDUCT ROLES

Every exercise activity should include the use of controllers, evaluators, and simulators. Each has a unique role and responsibility. The exercise scope, objectives, and budget should dictate the number and mix.

Robert Bradshaw
President
Bradshaw Emergency Management

This chapter is designed to help you understand the unique roles of controllers, evaluators, and simulators in supporting the conduct of a drill or exercise.

The objectives of this chapter are to

- train you in each of the controller, evaluator, and simulator roles;
- illustrate the differences between the roles.

Roles Of Controllers, Evaluators, And Simulators

You now know that to conduct a drill or an exercise you need to staff your conduct organization with controllers, evaluators, and simulators. To support the conduct of a tabletop exercise, you need facilitators and evaluators. In this chapter you will learn in greater detail what each of those positions entails.

Purpose of Designating Positions

The purpose of assigning personnel to the different conduct roles is to ensure that, through proper placement, the activity will be effectively managed, controlled or facilitated, and evaluated. Your goal is to have the activity result in a positive learning opportunity for participants. Additionally, the guidelines on combining roles and defining communication requirements among positions must be well understood to plan and implement conduct properly.

Review of Positions

Controllers, evaluators, and simulators are the people responsible for conducting a drill or exercise. Facilitators and evaluators support tabletop exercises. Observers and players can be found in all types of exercising activities. Remember, there does not exist a single or common set of terms when discussing exercising. However, while the positions may be described by different names, the responsibilities are basically the same across industries and jurisdictions.

The general positions include

- Coordinator (sometimes referred to as Director)
- Controller
- Evaluator
- Simulator
- Facilitator
- Player
- Observer

In addition, when drill or exercise activities are large in scale or scope, it is necessary to designate lead positions. These can include

- Simulation Cell Lead(s)
- Lead Controller(s)
- Lead Evaluator(s)
- Functional Lead Controllers/Evaluators

Full descriptions of these positions follow and definitions are contained in the glossary of the book.

The Coordinator: The coordinator has overall responsibility for conduct. In this capacity, the coordinator is in charge of all activities and personnel and will ensure conduct proceeds smoothly and according to the scenario.

Controller and Evaluator Roles and Responsibilities

Controllers are responsible for ensuring that the scenario proceeds as planned, that player activities remain within the design parameters of the drill, and that players have ample opportunity to address drill objectives. Evaluators are responsible for observing and documenting player decisions and actions. They also evaluate the effectiveness of those decisions and response actions based upon crisis and emergency plans and procedures, defined objectives, and evaluation criteria.

Specific responsibilities of controllers/evaluators (as a combined role) are to

Pre-conduct

- be familiar with the assigned organization's crisis and emergency management roles and responsibilities to be controlled and evaluated during conduct;
- participate in controller/evaluator briefings and support the scenario "dry run" as requested;
- review and be familiar with the scenario, objectives, and other guidance materials in the handbook prior to conduct;

During conduct

- monitor player activities in light of established ground rules for play;
- assess the progress of the scenario and the pace of activity to ensure that the scenario remains on track;
- provide frequent feedback to the simulation cell lead on conduct activities and implications for the unfolding scenario;
- coordinate needed adjustments in the scenario directly with the simulation cell lead;
- observe receipt of scenario information by players and evaluate their ability to respond effectively;

- maintain an observation log of player activities and record observations of, and any recommendations for, player performance as it pertains to identified objectives and sound crisis/emergency management practices throughout conduct;
- coordinate with other controller/evaluators and the simulation cell lead should any problems or questions arise concerning the bounds of play or the adequacy and content of message inputs;

Post-conduct

- conduct, participate in, and document a debriefing session for players at their assigned location immediately following termination;
- gather and label all player materials following termination of conduct activities;
- complete all evaluation forms;
- participate in a formal debriefing session among the conduct organization.

Simulator Roles and Responsibilities

Simulators are individuals acting on behalf of any person, agency, or service that would normally interact with your organization during an emergency or crisis situation. Simulators play these "roles" in order to provide realistic scenario message input that controls the flow of the scenario and presents the postulated problems and expected reactions by the organization's numerous stakeholders.

Specific responsibilities of simulators are to

Pre-conduct

- be familiar with the crisis and emergency response capabilities of the particular areas with which they are likely to interact;
- participate in simulator training and a scenario "dry run";
- review and become familiar with the scenario and assigned role described in the handbook prior to conduct;

During conduct

- input pre-scripted messages to drive the scenario (via telephone, computer, fax, other, as directed);
- respond to requests for information from players within designed/realistic response parameters;
- maintain a log of all communications and complete all forms for each communication;
- coordinate with the simulation cell lead in using status boards to track problems and open items;

- coordinate in advance with the simulation cell lead on any ad hoc or additional scenario inputs and accept any new role assignments from the simulation cell lead;

Post-conduct

- participate with the entire simulation cell in a simulator debriefing session covering an assessment of player performance and overall simulation activities;
- gather, label, and provide all conduct materials to the simulation cell assistant at termination and complete any evaluation forms, as instructed.

Simulation Cell Lead: The simulation cell is supervised by a simulation cell lead. Simulators should take direction from the simulation cell lead. Problems or questions should be directed to this individual. The simulation cell lead will periodically work with small groups of simulators to adjust the scenario or inject needed information. All additional tasks and requests from the simulation cell lead should be accepted and implemented.

Facilitator: Facilitators are involved in tabletop exercises. They are responsible for leading the group discussion so that all relevant issues and questions from the scenario are raised and explored. Facilitators are also responsible for leading the action planning session where participants identify action items.

Observer: Observers can be involved in both drills and exercises as well as tabletop exercises. Observers are invited guests who have no official role in the conduct of the exercise activity, although they are sometimes asked to submit their observations and insights from the activity. They are typically invited in order to encourage shared learning opportunities but have no formal responsibilities for conduct.

Player: A player is any individual or member of the response organization who is participating in a defined response role during the exercise activity. Players are responsible for performing their response roles as defined by plans and procedures and for participating in a realistic and professional manner.

In an effort to make your exercise activity more cost-effective, you may wish to assign personnel to more than one role. Though not recommended in larger drills and exercises, it may prove feasible in the smaller-scale activities.

The roles that can be combined fairly effectively are (1) the controller and evaluator positions in drills and exercises, (2) the simulation cell lead with simulator role in drills and exercises, and (3) the assistant facilitator with the evaluator role in tabletop exercises.

There are a number of advantages to combining roles. It can be economically efficient, thus allowing you to spread your limited resources further. It is better to have combined controller/evaluators in more sites than to have one each in fewer sites.

There are also a number of disadvantages to combining roles. In large, complex exercises it can minimize effectiveness of either or both roles. For example, it is difficult for personnel to control play as well as conduct a thorough evaluation. Combining roles may overextend resources and reduce the value you receive from your exercise activity.

For drills or exercises, you should rarely combine simulator with controller roles, and never combine simulation cell lead with controller roles. For tabletop exercises, you should never combine facilitator with evaluator roles. Observers of all types of activity should not be combined with any role. The very nature of several of these roles and their responsibilities tends to preclude them from being combined.

Communication Among Roles: Regular communications among the "leads" (i.e., simulation cell lead, lead controller, lead evaluator) during a drill or exercise is needed

- when there is an issue or concern during play,
- to provide feedback on how response is unfolding,
- to help control play and the scenario flow,
- to help in assessment of integration,
- to ensure all objectives have the opportunity to be demonstrated.

Controllers provide feedback to the simulators on how the scenario is being played so that simulators can make adjustments to the scenario, as needed. The "leads" and evaluators are better able to assess integration and information flow by communicating with each other, which gives them a better overall picture of player performance.

The controllers, evaluators, and simulators are the backbone of the conduct phase for drills and exercises. The same is true for facilitators during conduct of a tabletop exercise. Without qualified personnel in these positions, you can lose value from your exercise activity.

When selecting personnel to support the conduct of your exercise activity, it is important to refer back to the roles and responsibilities of each position. It will also be essential for you to conduct a briefing or training session for all selected individuals, as described in the development and conduct chapters of this book.

CHAPTER 8

Capturing Results Through Evaluation

The evaluation and action steps following the exercise are perhaps the most critical but often do not receive the attention they warrant.

Walter Scott
Professor of Management
Kellogg School of Management
Northwestern University
Senior Austin Fellow

This chapter highlights ways to document players' performance during exercise activities and compare them to expected responses, objectives of the exercise activity, and performance criteria.

The objectives are to help you

- identify the "why," "when," and "how" of evaluating performance in exercise activities;
- learn a process that feeds both the results and ways to improve into the continuous improvement effort of your emergency and crisis management program.

Introduction

The evaluation phase of the exercising process documents player responses and activities and compares performance to expected actions, objectives of the exercise activity, and established performance criteria. Personnel specially selected to perform as evaluators accomplish the evaluation activities. The evaluation phase is depicted in Figure 8-1.

Purpose

A tabletop exercise, drill, or exercise without effective evaluation or feedback offers limited value except for providing personnel with an opportunity to perform in their roles. Evaluation offers the means to

- collect data, analyze, and document how well players understand and perform their designated response roles;

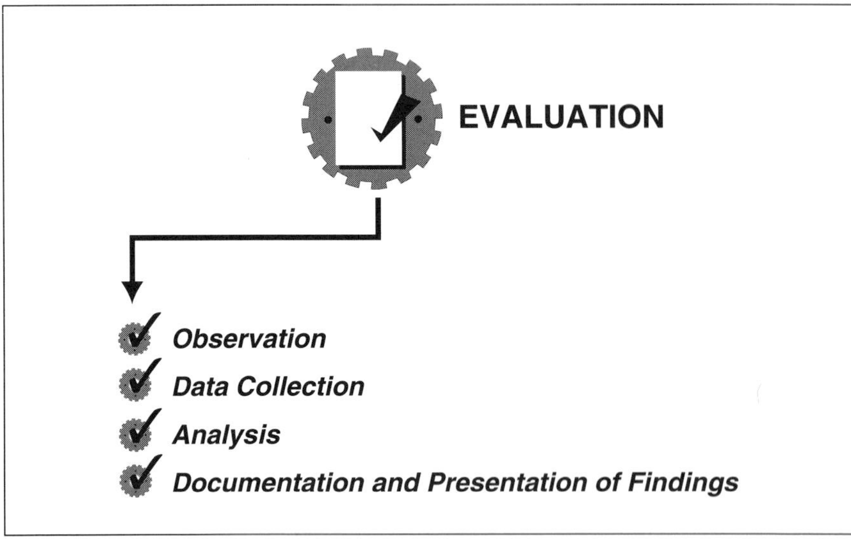

Figure 8-1
The Evaluation Phase

- compare player performance to expected responses, objectives, and performance criteria;
- assess the consolidated capabilities of teams and organizations;
- ensure the continual improvement of emergency and crisis response capabilities by documenting strengths and areas for improvement.

Evaluation Phase

This phase begins with the basic design and preparation for evaluation, continues through exercise conduct, includes the debriefing activities, and is concluded with the completion and presentation of a final evaluation. Evaluation of a tabletop exercise, drill, or exercise follows largely the same process outlined by the evaluation components listed below, although the number of evaluators and extent of evaluation are much greater for a drill or exercise. The components of evaluation and the evaluator requirements are described below. These form the basis of your evaluation plan. Specific evaluation considerations are then provided for tabletop exercises and drills/exercises.

Components of Evaluation

In order to plan the evaluation for an exercise activity effectively, it is necessary to consider the concepts of "what" will be evaluated and "how" to evaluate. These concepts were begun during the design and development phase. The actual plan for evaluation builds on the questions you asked and began to answer during these earlier phases. The question of "what" will be evaluated is based largely on the objectives, issues, and extent and scope of play. These outline what will be demonstrated and accomplished during the exercise activity. The "how" to evaluate manifests itself primarily in the four components of evaluation described below. It is important to note, however, that the selection and training of evaluators directly impact the quality and extent of your evaluation. Evaluator selection is addressed later in this chapter.

Step 1: Observation and Documentation

The first step in actually performing an evaluation is to observe players and document these observations. During conduct, the evaluator(s) must quickly document player and group responses and activities, as appropriate. To assist in this, evaluation forms are typically created that clearly reflect the exercise activity evaluation points and provide criteria (where possible) for evaluation of performance. These forms will assist the evaluator(s) in making observations and focusing assessment efforts. Documentation of the exercise activity using these forms should be in an appropriate level of detail and format to match the type of exercise activity and to facilitate reliable analysis of performance (drills or exercises) or discussion (tabletop exercises). The evaluator(s) should have a working knowledge of the response structure, plans, procedures, job aids, and checklists. Evaluator observations should identify both areas for improvement and positive or effective responses that deserve credit and might serve as reinforcement and learning tools for the rest of the organization. Positive feedback and reinforcement are critical elements of an effective evaluation process. To assist in designing appropriate evaluation forms, ask these questions:

- How will effective performance be evaluated?
- What decisions and/or actions should scenario events lead to?
- What key responses need to be observed?

Evaluators frequently use other tools, such as observation logs, to facilitate documentation of exercise activities. These logs are used to document player responses and to capture immediate assessments and evaluator observations as the activity unfolds. An observation log is shown in Figure 8-2. Written player evaluations are also usually captured on forms, like the one shown in Figure 8-3.

Step 2: Data Collection

Several types of data should be collected during and after the exercise activity. In addition to the notes included in the completed evaluation forms discussed in Step 1, other information such as data from players, work products of players, and observations from simulators (if applicable) should be assembled. Debriefings or action planning sessions for players and similar sessions with the conduct group, (i.e., evaluators, simulators and controllers for drills/exercises, and facilitator(s) and evaluator(s) for tabletop exercises) will also provide important data. Debriefings of players will often provide the most important evaluation observations. Data collection should address not only observation of player performance, but also provide information on the effectiveness of exercise design, development, and conduct. To prepare for this step effectively, ask yourself these questions:

- How have you planned to receive player input and self-evaluation?
- How will you arrange for the gathering of player response materials, logs, and forms?
- How will specific feedback on the effectiveness of the exercise or drill be captured from appropriate groups/individuals?

Observation Log Form

Page____ of____

Name: ____________________ Assignment: ______________________________ Date: __________

TIME	KEY ITEM	ASSIGNMENT LOCATION	OBSERVATION / EVENT DESCRIPTION / ASSESSMENT

Figure 8-2
Observation Log Form

PLAYER EVALUATION FORM

Please take a few minutes to complete this form. Please use additional sheets of paper for more room and return it to a controller/evaluator during the debriefing session after the exercise or as soon as possible.

What changes or additions to the current plans and procedures should be made based on your participation in the drill?

What additional job aids, checklists, etc., would you like to have made available?

Based on the drill, what additional capabilities/equipment/resources are needed?

What recommendation do you have to improve future drills and how should lessons learned be shared throughout the organization?

NAME AND ADDRESS: ______________________________

NOTE: *This information will be kept confidential, but we may need to contact you for further information.*

Figure 8-3
Player Evaluation Form

Step 3: Data Analysis

This step links the objectives of the exercise activity to the actual decisions and response actions of players. Most exercise activities do not require a detailed reconstruction and analysis of minute-by-minute exercise activities. The analysis should, however, include sufficient detail to support the observations of performance made by both evaluators and players. Planning for analysis should include activities to evaluate how effectively the design and development allowed accomplishment of exercise objectives. To help you with analysis of data, ask these questions:

- Were the objectives of the exercise met and expected player performance demonstrated?
- Was the response (and/or discussions) carried out effectively? According to plan?
- Would the response meet internal quality standards as well as the expectations of regulators and other critical external groups?
- How could performance be improved?
- Did the exercise provide a realistic learning opportunity for players?
- How effectively did the design and development allow successful demonstration of exercise objectives?

Step 4: Documentation of Evaluation

Goals for this step are twofold—to develop a format and presentation that meets the feedback needs of management and players while appropriately representing the consensus of the evaluation team. While there will be occasions where management and participants do not desire a written document, it is generally desirable to provide some type of written report to facilitate continuous improvement. Written documentation is also typically required to satisfy regulatory requirements. In this case, formats for the evaluation may be dictated by the agency. For extensive exercises, there may be a need to segment the evaluation into several evaluation products, each focusing on different aspects of exercise performance. Wherever this is the case, care should be given to avoid variation in evaluation reporting results. As noted in Step 1, provisions should be made to document both positive lessons learned and areas for improvement. The evaluation format should also allow participants the opportunity to comment and request clarification. To help you with this step, ask these questions:

- Is there going to be informal feedback? Verbal or written?
- Is there going to be a formal written report? Who is the audience and what level of detail is appropriate?
- Will a presentation of findings be conducted?

- How will reports and briefings on exercise evaluation be used?
- Is there a need for more than one means of feedback?

Planning for the evaluation of exercise activities should recognize that *HOW* it is accomplished may vary dramatically depending on the objectives of the exercise and the type of exercise activity being conducted. The four steps of the evaluation process, coupled with the evaluator activities described below, must be considered together when planning evaluation.

Evaluator Selection and Expectations

The evaluation plan is also built upon the basic functions expected to be performed by evaluators. Selection of qualified, skilled evaluators directly relates to the quality of the final evaluation. Qualifications can vary and may be defined in many ways. For example, response experience is helpful, and individuals selected as evaluators should be largely independent and provide a "third party" assessment. In addition to meeting selection criteria or qualifications, evaluators should be required to participate in several activities to ensure pertinent information is collected and assimilated for use in completing the evaluation process. These activities typically include

- participation in an evaluator training session and exercise activity briefing prior to the exercise activity;
- evaluation during conduct, including the "art" of being in the right location(s) at the right time(s) to ensure proper observation;
- participation in player and controller/evaluator debriefing sessions following conduct, including one-on-one interaction with select individuals to resolve evaluation issues;
- completion and submittal of evaluation forms and logs and performance assessments;
- provision of additional comments or clarification, if required.

Evaluators should be required to provide detailed notes and descriptions supporting evaluation points. Success in the evaluation is achieved to the extent that it reflects existing capabilities and provides direction for continuing activities to improve the organization's capabilities. The following outlines the differences that exist between the evaluation of a tabletop exercise and that of a drill or exercise.

Plan for *Tabletop Exercise* Evaluation

In a tabletop exercise evaluation, one individual will typically have primary responsibility for all stages of evaluation with input added by the facilitator. As previously noted, the evaluator in a tabletop exercise has also usually been fully involved in the design and development phases. He or she will therefore have excellent insight into the expected outcomes and desired discussion issues and points.

Observation logs should be used, and a limited number of evaluation forms may be needed. Due to the nature of tabletop exercises, a single evaluator can observe all player discussions during the session and will hear the feedback and results of the action planning/debriefing session directly from the players. Later, the facilitator(s) and evaluator should debrief and consolidate observations and appropriate recommendations. An evaluation report is usually written and typically ranges from 5 to 10 pages in length. Much of the content is obtained from, and reflects, the tabletop exercise action planning or debriefing session(s).

Plan for *Drill or Exercise* Evaluation

In designing and developing your drill or exercise, you must consider the *HOW* of evaluation. Effective planning for evaluation should include the development of appropriate forms, instructions, and training that provide evaluators with a common understanding of the *HOW*.

To aid the evaluation process, care should be given to the development of supporting evaluation forms. The forms should be developed around the previously developed exercise objectives and the desired extent and scope of play. Questions and appropriate criteria related to each expected performance area should be included to guide the evaluation and in fact may be supplemented or dictated by regulation or industry standards. Specific forms should be developed for the evaluation of player understanding of skill in emergency response or crisis management, in addition to that of individual and team roles and responsibilities. These evaluations should also reflect performance in relation to existing plans and procedures. Forms should call for an assessment of both strengths and areas for improvement observed during the activity. Separate forms should be provided for the evaluators' suggestions on how the current crisis and emergency management capabilities might be enhanced and for suggestions as to how the organization might build on the exercise in future exercise activities. The following are samples of two types of evaluation forms.

Sample

Evaluator Evaluation Form General CM Program

The overall purpose of the exercise is to demonstrate and validate your group's integrated crisis management capability and to reinforce and enhance these capabilities throughout all levels of the organization. To assist in measuring how well your group's crisis management structure performed as a total unit, evaluators are requested to provide general evaluation comments on this form.

Overall Observations Concerning Management Response (include at least three positive observations):

____
____
____
____
____
____
____
____
____
____
____

Recommendations for Improvements to the Response Structure, Planning Documents, Training, or Other Areas:

____
____
____
____
____
____
____
____
____
____
____
____

Sample

Evaluator Evaluation Form CM Prog-1

Objective: To validate the current status of the Crisis Management Program by demonstrating use and integration of appropriate plans and procedures.

Criteria: To satisfy this objective, participants within the CM structure should demonstrate sufficient knowledge of the CM program through the use of checklists and development of response options in a manner that exhibits a thorough familiarization of plans and procedures. Evaluators should also note the setup and functional capability of work areas where teams operate.

EVALUATION QUESTIONS:

Was there a general comprehension of individual and overall corporate structures and plans?

Were plans and procedures used? Did they facilitate coordination and information flow throughout all responding groups (i.e., integrated response)? Describe.

Were rooms/work areas functionally organized?

Is the demonstrated capability adequate? Is it adequately reflected in planning documents?

Was this objective accomplished? (explain)

Provide detailed observations:

Recommendations:

Instructions to evaluators may be provided in written form, either as a separate document or included on the evaluation forms. Evaluation will often require that instructions on how to evaluate specific activities at certain locations be provided to individual evaluators, in addition to any general instructions. While an experienced evaluation team may require only limited instructions, more extensive instructions may be advantageous where the team is less experienced or where exercise activities are expected to be complex or intense. The following is a sample evaluator instruction form.

Sample

Evaluation Instructions

Evaluators will be required to participate in an evaluation process that will involve several steps to ensure pertinent information is collected for use in completing an evaluation report. The steps include

- participation in training session prior to the exercise;
- participation in player and controller/evaluator debriefing sessions;
- completion and submittal of completed, detailed evaluation forms and logs;
- provision of additional comments or clarification, if required.

Evaluators will be required to provide detailed notes and descriptions supporting evaluation points using the forms provided here. Make sure to identify both areas for improvement and performance systems or functions that deserve credit and/or could be used as models for the rest of the organization. Using your knowledge gained during evaluator briefings and with information and forms in the handbook, provide input, keeping in mind what should have happened versus what did happen.

1. Consider the extent to which your group satisfied the general and specific objectives listed on the following pages. In addition to answering the specific questions on each evaluation form, identify
    - whether the objective was met/not met/partially met;
    - how it was accomplished or why it was not accomplished;
    - what factors affected performance of the objective;
    - changes in plans/procedures/staffing that will facilitate improved capabilities;
    - additional training/exercises to further develop capabilities.

2. As you observe and evaluate your specific functional or other assigned area, document in your observation log the following:
   - whether key issues were raised,
   - how each was/was not addressed and resolved,
   - whether each was managed in a timely and effective manner.

3 Finally, identify specific observations of performance not covered by the objectives/issues listed above. List at least three positive observations on exercise evaluation forms about what worked well, and identify any other places where and how capabilities could be improved. Specifically, consider a review and assessment of the following:
   - organizational structure,
   - plans and procedures (effectiveness and overall soundness),
   - staffing (demonstrated skills/knowledge and suggested staffing changes),
   - equipment and facilities,
   - information flow and coordination.

Evaluator training provides an important opportunity to assure a full understanding by evaluators of *HOW* they are expected to accomplish evaluation. Even in the rare circumstance where you can use an evaluation team that consistently works together on a regular basis, some formal training on the provisions of the scenario and the evaluation expectations will be necessary. However, it is typical that training should be more extensive, to include simulated evaluation opportunities that allow evaluators to observe and demonstrate how actual evaluation will be completed and documented.

Throughout the design and development of the drill or exercise you should consider a number of *HOW* questions—inserting the answers into the forms, instructions, and training provided to the evaluation team.

To consider how evaluators will *collect data*, ask these questions:

- Will debriefing sessions be conducted for players immediately following the conclusion of the exercise?
- Will debriefings sessions be held for the conduct organization (controllers/simulators/evaluators) on player performance and administration of the drill or exercise?
- What process will be used to retrieve and compile all job aids, written notes, or other relevant documentation from players as well as the exercise organization?
- How will the evaluators' observations be captured?

To consider how evaluators will *analyze data and performance*, ask these questions:

- How will it be determined if identified objectives of the exercise were met?
- How will observations and recommendations comparing player response to performance criteria be documented?
- How will the positive, proactive response capabilities, as well as areas for improvement, be recognized and documented?
- How will effectiveness and efficiency of operations be evaluated?
- Will specific recommendations be made for each area requiring improvement?
- Will general organization-wide recommendations be made for each area requiring improvement?
- How will feedback on the design of the exercise activity be obtained from players and the conduct organization?

To consider how findings and recommendations for improvement will be *documented*, ask these questions:

- Does the process ensure player and evaluator recommendations are considered and/or incorporated?
- Does the process include the evaluation team and the coordinator in the review and approval of the final draft report and/or presentation?
- Does the process ensure that any written report fairly represents the consensus of the evaluation team and coordinator?
- Does the process ensure that players receive detailed and complete feedback?
- Does the process ensure that management has an opportunity to comment and request clarification where necessary?

To consider how the evaluation will be *presented*, ask these questions:

- What schedule is appropriate for disseminating exercise activity results to management and players?
- How best can results be presented and disseminated to the various audiences?
- How will the findings be incorporated into the program tracking system to monitor actions taken with regard to improvements?

In the evaluation of a drill or exercise, the responsibility is divided among multiple evaluators who are each accountable for a specific segment of the evaluation. The final evaluation must tie together input from all evaluators and must address the extent to which communication and interaction among various segments was successful. Each evaluator should solicit input from the players as well as controllers and simulators for their particular segment of the drill or exercise. The evaluation team, or the core evaluation group, must integrate the collective input, reconcile any discrepancies, and draw appropriate conclusions. The length of the evaluation report is typically 20 to 200+ pages, depending upon the complexity and size of the drill or exercise and level of detail required.

While the specifics of *HOW* do and should vary, the quality of evaluation is measured in its use as a means to identify lessons learned and to guide continuous improvement of the emergency and crisis management capabilities. The next step is to feed the lessons learned into your tracking system and to achieve continuous improvement.

HOW TO

Plan for Evaluation

- ☐ *Assess Evaluation Needs*
  - *Player response locations*
  - *Targeted performance for evaluation*
- ☐ *Identify Audience(s) for Evaluation Feedback*
- ☐ *Determine Form(s) and Format(s) of Evaluation Feedback*
  - *Verbal debriefings*
  - *Detailed evaluation report*
  - *Evaluation presentation*
  - *Evaluation highlight report*
- ☐ *Identify Activities and Tools Needed to Support Effective Evaluations*
  - *Review expected play*
  - *Identify information/data collection needs*
  - *Develop needed materials and tools*
  - *Prepare for designated activities*
- ☐ *Select Qualified Evaluators*
- ☐ *Make Evaluation Assignments*
- ☐ *Establish Your Exercise Schedule of Activities*
- ☐ *Create Your Evaluation Plan*

CHAPTER 9

Tracking Continuous Improvements

Tracking of corrective actions identified as a result of exercise activities is an important function. As such, it requires a defined process, clearly assigned responsibility, and follow through.

Glenn Epler
Commander
United States Coast Guard Former Chief
Pollution Response Branch

This chapter emphasizes and presents methods for tracking and monitoring the results of your exercise activities.

The objectives are to help you

- understand the importance of tracking exercise activity results and action items,
- understand the methods available to track results and action items.

Introduction

You have just completed your evaluation of your exercise activity and have developed a list of action items and/or recommendations that need to be completed to improve your group's crisis and/or emergency management capabilities. To get the most from every exercise activity, you must ensure that these issues do not get set aside without being settled or improvements implemented, or you will have lost a valuable opportunity to improve your capabilities. To aid you in resolving these open issues and improvement needs in a timely manner, it is important to develop a tracking system. This tracking system can also be used to track any action items identified during a real event or other program activities.

Purpose of Tracking

The purpose of tracking lessons learned goes beyond the question of validation of capabilities; it answers the question, "How can we make it better?" It accomplishes this by helping your group

- broaden the value of an exercise activity by sharing lessons learned throughout the organization,
- use the information gathered during the exercise activity to improve and build on your organization's overall crisis and emergency management capabilities,
- revise plans and procedures,
- close the gaps in response capabilities,
- develop objectives for future exercise activities.

Failure to follow up on these lessons learned can place an organization in the difficult position of trying to explain itself in the aftermath of a real response. Legal and regulatory considerations also influence the need to resolve issues uncovered during an exercise activity.

The Five Steps to Tracking

The steps involved in tracking allow you to document lessons learned and areas for improvement, to monitor progress, and to track completion or resolution of issues. This process applies regardless of the type of exercise activity employed. There are five specific steps you should accomplish when tracking action items resulting from an exercise activity or real event:

1. List lessons learned, resulting action items, and recommendations from the response.
2. Assign responsibility and completion dates.
3. Develop method/process to complete action items.
4. Develop method/process to track completion of action items.
5. Identify capabilities and areas for future exercise objectives.

Taking a step-by-step approach, the following describes the process of tracking your lessons learned and improvements from any exercise activity.

Step 1 - List all lessons learned and action items: You should already have completed a list of areas for improvement during the evaluation phase of the exercise activity or in your debriefing (and post-incident review) following a real event. If not, look at each aspect of the response demonstrated to determine if an improvement or modification is desired/needed. Review and reformat your list into an action item format, being careful not to rush to a solution unless a consensus was reached during conduct. In other words, state your action items as *what* needs improvement or modification, not *how* this will be accomplished.

Next, clearly distinguish between those actions that are merely desirable and those that are critical for an effective performance by the organization in a crisis. This will help you and your organization prioritize efforts.

Step 2 - Assign responsibility and completion dates: Designate a responsible party for each action item. This can be an individual or a group and should be based on the functional or oversight area associated with the action item. It is the responsibility of the designated party to identify solutions and take the actions needed to resolve the action item. Action items may also be bundled and assigned where the same area or function is involved.

It is important to assign a specific date as well as possible interim dates for achieving milestones in the implementation of solutions or improvements. The date provides a working timeline and should demonstrate a reasonable effort to take corrective action. Regulatory compliance and other factors, such as the critical nature of the improvement and likely complexity of the solution, should be considered when setting these dates.

Step 3 - Develop a method/process to complete action items: Different action items may require different methods to achieve resolution. While it is the responsibility of the designated party to choose the method of resolution, coordination with other designated parties is generally desirable. Some examples of improvement methods are

- plan revision;
- development of new plans, procedures, or response tools;
- increased/new training for personnel;
- increased reserve of resources;
- no action should be taken.

Step 4 - Develop method/process to track completion of action items: The status of open or unresolved action items should be reviewed periodically to ensure that corrective actions are being taken within the designated time frame in a progressive fashion.

There are several ways to track these action items so that you are aware of tasks which have been reviewed, tasks which are completed, and tasks which are open. Some examples are to

- request that designated parties submit oral/written reports on interim/final status;
- track status manually, using a responsibility matrix (see sample in Figure 9-1);
- develop standardized status forms (see sample in Figure 9-2);
- conduct follow-up meetings with designated parties;
- develop/use computer software tracking programs.

The method is less important than ensuring that all action items are tracked through resolution.

Step 5 - Identify capabilities and areas for future exercise objectives: As you begin to develop new or improved capabilities, it becomes important to validate them through another exercise activity. Therefore, you should have a process in place to incorporate these new capabilities into objectives for future exercises. To do this you should

- maintain an updated database of exercise objectives,
- add or emphasize objectives in scheduled exercise activities as new or improved capabilities become operational,
- incorporate these objectives into your multi-year exercise planning schedule.

The inclusion of a tracking process in your exercise program makes it an effective mechanism to support continuous improvement.

ACTION ITEMS	EMERGENCY RESPONSE COORDINATOR	HUMANITARIAN ASSISTANCE	FINANCE	CRISIS MANAGEMENT TEAM	PUBLIC AFFAIRS	INCIDENT COMMANDER	CORRECTIVE ACTIONS
EXAMPLE: Increase communications between response organizations	***D***			***S***		***C***	Improve communication channel and process between response organizations
EXAMPLE: Respond to media inquires with a common message					***D***		Revise P&GA information gathering mechanism and enhance approval process
EXAMPLE: Improve death notification process	***O***	***D***					Conduct H.A. training for designated personnel

KEY:
D = *Direct Responsibility*
S = *Supporting Role*
O = *Oversight Responsibility*
C = *Coordination Role*

Figure 9-1
Responsibility Matrix

TRACKING STATUS FORM

Action Item	Assigned To:	Corrective Actions:	Milestones	Completion Date

Instructions:
1) List all action items
2) Assign responsibility
3) Describe the agreed-upon corrective actions
4) Assign a completion date and relevant milestone dates

Figure 9-2
Tracking Status Form

HOW TO

Track Continuous Improvement

- ☐ *Create Action Item List*
  - *Lessons learned*
  - *Recommended improvements*
  - *Identified action items*
- ☐ *Designate Responsible Party*
  - *Individual*
  - *Team*
- ☐ *Assign Responsibilities and Deadlines*
  - *Individual action item*
  - *Bundled action items*
  - *Deadlines for completion*
- ☐ *Track Proposed Solutions and Completion of Improvements*
  - *Plan revisions*
  - *New planning document development*
  - *Training enhancements*
  - *Enhanced resources*
  - *No action required*
- ☐ *Use Tracking Tools*
  - *Update (verbal/written)*
  - *Responsibility matrix*
  - *Computer software*

CHAPTER 10

Closing The Loop: Linking Back To Your Program

It is important to keep in mind that simulation exercises are a means to an end; their ultimate goal is the continuous improvement of both crisis prevention and crisis management capability.

Mary Woodell
Senior Principal
CMRC & Co.

This chapter is designed to help you understand how the results of exercise activities link to other elements of your crisis and emergency management program and how these results support the concept of continuous improvement.

The objectives are to help you

- better understand the relationship of exercise activities to the rest of your crisis management program,
- recognize the value of results from exercises.

Introduction

This final chapter is designed to refocus your mind once again on the bigger picture, remembering that the purpose of exercising is to validate and enhance capability. These goals are best achieved when the exercise lessons learned and the ideas for improvement are linked into the crisis and emergency management program to be shared throughout the organization.

Purpose

The purpose of continuous improvement is to ensure that the lessons learned and corrective actions and areas for improvement identified in exercise activities are

- integrated into the prevention, planning, training, quality assurance, and future exercise considerations of the organization;
- properly shared to maximize value throughout the organization.

How to Integrate Your Exercise Evaluation into Your Crisis and Emergency Management Program

As we have discussed throughout this book, exercises enable organizations to validate capabilities and identify areas for improvement. The process of identifying areas for improvement is important; however, it is equally important that these improvements be linked and incorporated into existing procedures and programs. These linkages serve as the vehicle to enhance capability. To illustrate ways in which exercising can be integrated, each program element is described here. Methods to link them to exercising results are also discussed.

1. Identify opportunities for prevention: Exercising activities often lead to prevention opportunities. An exercise participant may observe that a particular system or process could be modified to decrease the chances of the occurrence of an incident. Questions you may ask include:

- Did the exercise activity highlight a risk or vulnerability that could be reduced if modifications were made to systems, storage quantity location, or handling of materials, etc.?
- Did the evaluation indicate that if it had been a real-life situation, the current warning system would not have alerted personnel until it was too late?
- Did the scenario outline a series of consequences so severe that a business area, operation, or facility is deemed too "risky" to continue?

2. Incorporate suggested improvemnts into planning and related documents where appropriate: Response activities and the use of planning documents and related materials during an exercise will usually lead to suggested improvements for those documents. Modifications may range from increasing the font size used for phone numbers to adding roles or responsibilities undefined. Questions you may ask include:

- Were there suggestions to modify any roles and responsibilities?
- What sections of the plan or procedures seemed most helpful to responders?
- Were the plans/procedures in a useful and useable format?
- What additional information would make these documents more useful?
- Would additional or modified tools (e.g., status boards, checklist, etc.) be useful to facilitate effective response?
- What plan changes and modifications would appropriately reflect the current concept of operations?

3. Modify and add training activities as appropriate: Additional training requirements may be highlighted as a result of an exercise or due to resulting modifications made to plans and procedures. For example, if an exercise leads to the identification of new or modified roles and responsibilities, affected personnel should receive related training. Questions you may ask include:

- Were personnel adequately trained to perform the roles and responsibilities required of them?
- Were changes in roles or processes suggested as a result of the exercise?
- Were additional positions determined to be necessary?

4. Integrate exercise activity lessons learned into the quality assurance efforts: Linking exercise evaluations to quality assurance (QA) efforts is essential because exercises are a measure of performance. Expected levels of performance or performance standards may be increased based upon lessons learned during the exercise activity. For example, prior to an exercise, the ability to be on-scene within 24 hours may be the established performance standard. The exercise might determine that it was essential for responders to be on-scene within just 12 hours. This new requirement should be integrated into the quality assurance programs, in addition to procedures. Questions you may ask include:

- Were new or modified performance requirements identified?
- Did the exercise provide any insight as to how to evaluate performance effectively and accurately?

5. Ensure evaluations are integrated with the post-incident review (PIR) process: Exercising and PIRs provide the best means available to an organization to assess capability. These activities should be linked to ensure minimal duplication in assessing the same capabilities. If an actual response meets identified exercise activity objectives, the planned activity can be redesigned to address different objectives. Questions you may ask include:

- Are lessons learned and areas for improvement shared between the PIR process and the exercise process?
- How are lessons learned from real events captured, shared, and built into your program efforts?
- Are exercise activities and real events being evaluated in a consistent manner?
- Are activities to improve tracking and sharing of lessons learned used consistently for both exercise activities and real events?

6. Consider areas for improvement as future exercise activity objectives: Evaluation results will almost certainly lead to modifications in one or more program areas. These modifications and the improved capabilities should be validated at some point in the future through an exercise activity. The validation should ensure that the modification or improved capability effectively accomplishes its intended function. Questions you may ask include:

- Were areas essential to effective response identified, thus warranting revalidation in a future exercise? For example, if additional training of notification staff is identified as an area for improvement, a future exercise activity objective might be to validate the enhanced notification procedures and capabilities.
- Has sufficient time lapsed to allow for the implementation of improvements?

Sharing Lessons Learned

The results of exercise activities in the form of lessons learned and improvement opportunities should be shared throughout the organization.

This maximizes the value of every exercise activity by allowing others who were not involved to learn the same lessons. For example, actions taken to address prevention opportunities identified in an exercise can be shared, thus decreasing the risks for similar locations and operations throughout the organization. The result can also enhance cooperation and further sharing of lessons learned throughout the organization when these kinds of benefits become apparent.

What Will Integration Achieve?

Integration of program activities and results, as described above, is critical to gaining the most benefit from each exercise activity. The exercise will have less of an impact on improving and sustaining capability if an activity is performed and recommendations are not captured in writing or if lessons learned and corrective actions are not implemented and shared. A lack of follow-through to improve known gaps in your capability can also represent legal considerations. Close integration of your program activities provides the organization with maximum value from its exercise program.

Integration should lead to the following list of benefits:

- Increased ability to improve upon and sustain capability;
- Enhanced overall organizational capability;
- Identification of prevention opportunities;
- Increased cooperation and sharing of lessons learned throughout the organization;
- A more "seamless" (integrated) response structure, as improvements are shared and incorporated into various program elements (training, quality assurance);

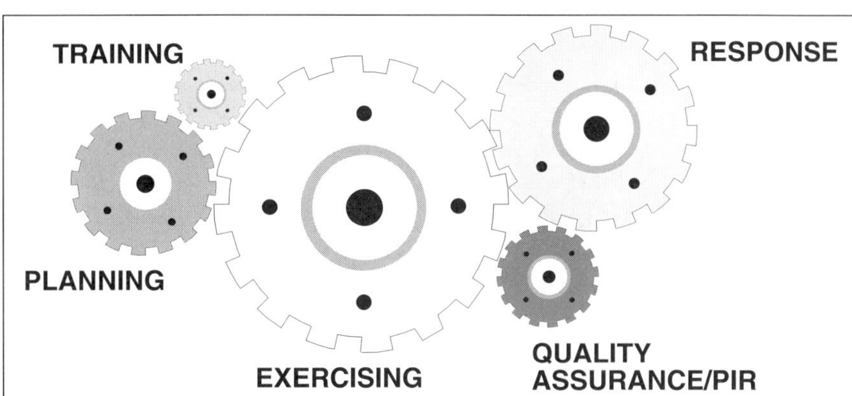

Figure 10-1
Crisis Management Program

- Decreased concern over legal considerations that may arise if areas for improvement are identified, but not acted upon;
- A high-quality, cost-effective exercise program.

Conclusion

As was stated previously, your exercise program supports the other activities in your crisis and emergency management program. Each activity represents a gear in the overall ability of the organization to respond to and manage a crisis effectively. Proper functioning of the system requires that all of the gears work together and turn at compatible speeds, as depicted in Figure 10-1. Integration of the various concepts and activities should lead to a continuous improvement process that benefits your crisis and emergency management program.

HOW TO

Close the Loop

- ☐ *List Action Items and Improvement Areas*
- ☐ *Link Improvements to Each Component of Your Program*
  - *- Prevention*
  - *- Planning*
  - *- Training*
  - *- Quality*
  - *- Post-incident reviews*
  - *- Future exercise activities*
- ☐ *Ensure Broad Sharing of Lessons Learned*
- ☐ *Track Program, Exercise Activity, and Improvements*
- ☐ *Achieve Maximum Value from Your Exercising Program!*

Glossary, Bibliography, Videotape References, and Index

Glossary

Artificiality: Conditions created by the design of an exercise that do not simulate or mirror actual conditions. May interfere with the player's ability to respond realistically.

Capability: The ability to perform with skill or knowledge, or provide a resource to meet a specific requirement.

Controller: An individual assigned to one or more exercise functions or locations to monitor and control play. Controllers monitor play to ensure that exercise participants adhere to established design parameters. They also observe, document, and initially evaluate exercise performance. Controllers are used in drills and exercises.

Crisis: A circumstance, event, or series of episodes that threatens to fundamentally affect or alter the way an organization conducts business.

Crisis Communication: The process of creating and executing a communications strategy that helps a corporation reach its operational objective and successfully avoid a crisis or minimize its impact.

Crisis Management: A proactive operating philosophy that advocates capabilities to prevent, prepare for, respond to, and recover from a circumstance, event, or series of episodes that threaten to fundamentally affect or alter the way an organization conducts its business.

Crisis Management Program: A program that provides for an integrated crisis management capability through planning, organizational development, training, exercising, and continuous improvements at each level of the organization.

Crisis Management Team (CMT): A group of senior executives or managers charged with developing and implementing crisis management efforts and strategies to minimize the impact of a crisis on the organization.

Critique: Also called a Debriefing. A meeting of players, facilitators and/or controllers, and evaluators following the conclusion of the exercise activity to provide essential comments on operations and performance during exercise play.

Disaster: The consequences of an incident, emergency, or crisis on the surrounding area.

Drill: An exercise activity requiring performance of response activities by individuals and teams, and limited mobilization of personnel and resources.

Emergency: An incident that threatens, based on impact/need to respond, human life, health, property, or the environment if not controlled, contained, and/or eliminated immediately.

Emergency Management: Those measures taken to prevent, prepare for, respond to, and recover from the acute effects of an emergency.

EOC: Emergency Operations Center, a facility or location from which parts of operational response and decision-making may be coordinated. These are often staffed with multi-organizational or multi-departmental representatives.

Exercise Organization: Those persons who are responsible for controlling, evaluating, or simulating exercise activity.

Evaluator: An individual assigned to one or more exercise functions or locations to document and evaluate individual, team, and organizational performance based on the exercise objectives and performance criteria. Evaluators are used in all types of exercise activities.

Exercise: An exercise activity requiring performance, integration, and coordination of response activities by several individuals and teams, as well as mobilization of personnel and resources.

Exercise Activity: An activity that provides an opportunity for participants to train in and practice emergency and crisis management skills. Exercise activities provide a method of evaluating participants' ability to meet emergency and crisis management requirements and responsibilities.

Exercise Coordinator: The individual responsible for the overall management and coordination of an exercise activity.

Exercise Management: The group of individuals responsible for administrating and coordinating the design, development, conduct, and evaluation of exercise activities.

Facilitator: A specially trained individual assigned responsibility for guiding participant discussions during tabletop exercises to ensure key issues are addressed.

Field Location: A geographic location, region, facility, or collection of field operations supported by a single emergency management organization.

Free Play: The policy of allowing players to respond as realistically as possible, within design parameters, without jeopardizing personnel or plant safety or expending unnecessary resources. Also applies to simulators responding to player responses.

Full-Scale Exercise: An exercise activity which validates the major aspects of an organization's crisis management program and involves all levels of the organization.

Hazard: Anything that has the potential for causing harm to the organizational operations, people who work at or live around the facilities, or for damaging property and the environment.

Hotwash Session: An immediate debriefing session between players and members of the control organization to discuss preliminary observations on exercise performance. Also see critique.

Humanitarian Assistance Program: An integrated program designed to provide prompt and appropriate assistance to employees, victims of and responders to a crisis event.

Incident: A set of circumstances or an event that affects normal operations, requires attention, and has the potential to precipitate an emergency or crisis.

Initial Conditions: A summary of existing conditions leading up to the start of the drill exercise. The summary sets the stage and is usually presented in written format to players before conduct begins.

Lead Controller: The person with overall responsibility for exercise management and information flow during drills and exercises. Decisions on deviations from the pre-scripted scenario or exercise termination are coordinated with this position.

Lead Evaluator: The person with overall responsibility for directing the documentation and evaluation of drills and exercises. This position reports to the exercise coordinator.

Master Scenario Events List: A full list or table which chronologically displays designated scenario time, event synopsis, expected response, and the objective to be demonstrated, if appropriate.

Message: The vehicle that provides information to prompt player response. Messages are disseminated by the exercise organization, usually simulators, and may be verbal, written, or in the form of a visual display.

Message Summary: A listing of all messages, including message numbers, times, sources, and a brief description of the content of each message.

Narrative Summary: A short overview of the exercise scenario written in paragraph form, outlining only the major events.

Objectives: The stated goals of exercise activities. Objectives define the level of skill and specific capabilities to be demonstrated by players during the exercise. Exercise objectives are used as the basis for evaluation of exercise performance or assessment of training effectiveness.

Performance Requirements: Those response activities required or expected of the organization, teams, or individuals, established by regulatory mandate, industry standard, or company policy.

Performance Standard: Criteria by which operational and management functions can be measured to evaluate the degree to which those functions have achieved a minimum level of quality.

Player: An individual or member of the response organization participating in the exercise activity in his or her role.

Real Time: When actual time is used for the simulated events to take place.

Scenario: A hypothetical situation or chain of events that depicts an incident, emergency, or crisis and all the associated consequences. Used to guide simulation during a drill or exercise.

Scenario Time: Scenario time is expressed in terms of time elapsed since the initiating event.

Simulation: An umbrella term that means an artificially produced condition that replicates a real-life emergency or crisis situation.

Simulation Cell: Sometimes called a Control Cell. The working location for those members of the exercise organization responsible for delivering messages and injecting information to prompt response actions by players.

Simulator: An individual assigned the responsibility to artificially duplicate (role-play) the response activities of personnel and groups not participating in the exercise.

Tabletop Exercise: An exercise activity conducted in a conference room setting involving discussion of a scenario by a response team.

Time Jump: A mechanism by which scenario events may be artificially accelerated in order to place players in situations that would occur at a future point in time. Time jumps require exercise play to be stopped and then to resume at some future point in time. Time jumps are done to include events that otherwise would not occur in the limited amount of time allowed for an exercise.

Timeline: A sequential listing of the times and key events in a scenario that drive player response.

Training: Activities undertaken to educate personnel assigned emergency response and crisis management roles and responsibilities. Designed to provide an opportunity to practice crisis and emergency management skills, ensuring that they are adequately prepared to fulfill these roles in the event of an incident, emergency, or crisis.

Trusted Agent: Individuals with unique or specialized expertise who are confidentially included in the scenario development to ensure realistic events are postulated and appropriate responses are anticipated.

Bibliography

Barton, Laurence. *Crisis in Organizations: Managing and Communicating in the Heat of Chaos*. Cincinnati, OH: South-Western Publishing, 1993.

Berge, Dieudonnee. *The First 24 Hours: A Comprehensive Guide to Successful Crisis Communications*. Cambridge, MA: Basil Blackwell, 1990.

Carroll, Paul. *Big Blues: The Unmaking of IBM*. New York, NY: Crown Publishers, Inc., 1993.

Chemical Emergency Preparedness Program Interim Guidance. U.S. Environmental Protection Agency, Office of Solid Waste and Emergency Response (WH-548A), Washington, D.C. 20460, 1985.

Comfort, Louise, et al. *From Crisis to Community: The Pittsburgh Oil Spill*. Amsterdam: Elsevier, 1989.

Community Awareness and Emergency Response Program Handbook. Chemical Manufacturers Association. Washington, D.C., 1985.

Coping with Crises: The Management of Disasters, Riots, and Terrorism. Edited by Uriel Rosenthal. Springfield, IL: CC Thomas, 1989.

Crisis Management: A Casebook. Edited by Michael T. Charles et al. Springfield, IL: CC Thomas, 1988.

Crisis Management Planning Manual for the Chemical Manufacturing Industry. Chemical Manufacturers Association, 2501 M Street, N.W., Washington, D.C. 20037, 1991.

Crisis Response: Inside Stories on Managing Image Under Siege. Edited by Jack A. Gottschalk. Detroit, MI: Visible Ink Press, 1993.

Davidson, Art. *In the Wake of the Exxon Valdez: The Devastating Impact of the Alaska Oil Spill*. San Francisco, CA: Sierra Club Books, 1990.

Developing a Hazardous Materials Exercise Program: A Handbook for State and Local Officials. National Response Team, September 1990.

Developing an Oil Spill Exercise Program. United States Coast Guard (USCG), 1993.

Ellis, Derek V. *Environments at Risk: Case Histories of Impact Assessment.* New York, NY: Springer-Verlag, 1989.

Emergency Response Guidebook, DOT P-5800.2. U.S. Department of Transportation, Research and Special Programs Administration, Materials Transportation Bureau, 1980.

Exercise Design Book: Exercise Scenarios. SM 170.3. Federal Emergency Management Agency, January 1989.

Exercise Design Book: Guide to Emergency Management Exercises. SM 170-2. Federal Emergency Management Agency, January 1989.

Exercising Emergency Plans Under Title III. SM 305.4. Federal Emergency Management Agency, September 1990.

Federal Preparedness Exercises. FPC-6. Federal Emergency Management Agency, April 1992.

Fink, Steven. *Crisis Management: Planning for the Inevitable.* AMA-COM, the American Management Association, 1986.

Gigliotti, Richard J. *Emergency Planning for Maximum Protection.* Boston, MA: Butterworth-Heinemann, 1991.

Ginn, Ronald D. *Continuity Planning: Preventing, Surviving, and Recovery from Disaster.* Oxford, England: Elsevier Advanced Technology, 1989.

Guide for the Development of State and Local Emergency Operations Plans, CPG 1-8. Federal Emergency Management Agency, Washington, D.C., 1992.

Guide to Exercises in Chemical Emergency Preparedness Programs. U.S. Environmental Protection Agency, May 1988.

Guidelines for Hazard Evaluation Procedures. American Institute of Chemical Engineers, Center for Chemical Plant Safety, 1985.

Hazardous Materials Exercise Evaluation Methodology. (HM-EEM) and Manual. Federal Emergency Management Agency, October 1989.

Hazardous Materials Emergency Planning Guide. National Response Team. March, 1987. Contains plan development and planning program guidance and recommendations.

Hazarika, Sanjoy. *Bhopal: The Lessons of a Tragedy*. New York, NY: Viking Penguin, 1987.

Janis, Irving L. *Crucial Decisions: Leadership in Policymaking and Crisis Management*. New York, NY: Free Press, 1989.

Keller, Maryann. *Collision: GM, Toyota, Volkswagen and the Race to Own the 21st Century*. New York, NY: Doubleday, 1993.

Lagadee, Patrick. *States of Emergency: Technological Failures and Social Destabilization*. Boston, MA: Butterworth-Heinemann, 1990.

Lindell, Michael K. *Behavioral Foundations of Community Emergency Planning*. Washington, D.C.: Hemisphere Publishing, 1992.

Managing Disaster: Strategies and Policy Perspectives. Edited by Louise K. Comfort. Durham, NC: Duke University Press, 1988.

Marconi, Joe. *Crisis Marketing: When Bad Things Happen to Good Companies*. Chicago, IL: Probus Publishing, 1992.

Margolick, David. *Undue Influence: The Epic Battle for the Johnson & Johnson Fortune*. New York, NY: William Morrow and Company, Inc., 1993.

Media, Crisis and Democracy: Mass Communication and the Disruption of Social Order. Edited by Marc Raboy. London, England: Sage Productions, 1992.

Meyers, Gerald C. with John Holusha. *When it Hits the Fan: Managing the Nine Crises of Business*. Boston, MA: Houghton Mifflin Company, 1986.

Mitroff, Ian and Christine M. Pearson. *Crisis Management: A Diagnostic Guide for Improving Your Organization's Crisis Preparedness*. San Francisco, CA: Jossey-Bass Publishers, July, 1993, p. 160.

Mitroff, Ian. *We're so Big and Powerful Nothing Bad can Happen to US; An Investigation of America's Crisis Prone Corporations*. Secaucus, NJ: Carol Publishing Group, 1990.

Mitroff, Ian. *Corporate Tragedies: Product Tampering, Sabotage, and Other Catastrophes*. New York, NY: Praeger, 1984.

Nothing to Fear: Risks and Hazards in American Society. Edited by Andrew Kirby, Tucson, AZ: University of Arizona Press, 1990.

NRT-1, Hazardous Materials Emergency Planning Guide. Federal Emergency Management Agency, 500 C St. S.W., Washington, D.C. 20472, 1987.

NRT1A, Criteria for Review of Hazardous Materials Emergency Plan. Federal Emergency Management Agency, 500 C St. S.W., Washington, D.C. 20472, 1988.

Nudell, Mayer. *The Handbook of Effective Emergency and Crisis Management*. Lexington, MA: Lexington Books, 1988.

On Scene Coordinator/Regional Response Team, Oil and Hazardous Substances Discharge Simulation, Planning Manual. U.S. Department of Transportation, United States Coast Guard.

Pauchant, Thierry C. *Transforming the Crisis-Prone Organization: Preventing Individual, Organizational, and Environmental Tragedies*. San Francisco, CA: Jossey-Bass Publishers, 1992.

Radiological Emergency Preparedness Exercise Manual. Federal Emergency Management Agency, November 1991.

Scenario Development Manual: Exercise Development. Federal Emergency Management Agency, 1985.

Schoenberg, Robert J. *The Art of Being a Boss: Inside Intelligence from Top-Level Business Leaders and Young Executives on the Move*. Philadelphia, PA, and New York, NY: J.B. Lippincott and Company, 1978. p. 260.

Scotti, Anthony J. *Executive Safety and International Terrorism: A Guide for Travellers*. Englewood Cliffs, NJ: Lexington Books, 1987.

Shiels, Frederick L. *Preventable Disasters: Why Governments Fail.* Savage, MD: Rowman and Littlefield, 1992.

Shrivastava, Paul. *Managing Industrial Crises: Lessons of Bhopal.* New Delhi, India: Vision Books, 1987.

Simulation Techniques for Hazardous Materials Exercises. Federal Emergency Management Agency, November 1991.

Site Emergency Response Planning. Chemical Manufacturers Association, Washington, D.C. 1986.

Trento, Susan B. *The Power House: Robert Keith Gray and the Selling of Access and Influence in Washington.* New York, NY: St. Martin's Press, 1992. p. 400.

Waugh, William L. *Terrorism and Emergency Management: Policy and Administration.* New York, NY: M. Dekker, 1990.

Videotape References

Alert 3. Tape of United Airlines flight 232 crash in Sioux City, Iowa. The first 13 minutes of this tape is actual audio from the pilot to air controllers. Also describes how they prepared for and responded to the crash, and preparation which began with an exercise two months prior. 50 minutes. Date: 1989. Source: G.R. Lindblade & Co., Sioux City, IA, (712) 255-4346.

American Heat (Session on United Airlines crash), Volume 4, Program 4. Gives an overview of events leading up to the crash. The preparation that Sioux City did during a drill helped with their rapid response and the saving of lives. May be better than tape above. Length of segment: 12 minutes. Source: American Heat Video Productions, St. Louis, MO, (800) 722-2443.

Amoco Corporation: Limited Exercise 1993. Documentary on Amoco's April 1993 Limited Exercise. Provides background on Amoco's crisis management program and the exercise design process. 10 minutes. Date: 1993. Available for viewing for educational purposes with books offered by the University of Houston-Clear Lake.

BP International: Exercise European Alert. Documentary on BP International's Exercise European Alert involving BP responders from a number of countries. Good example of scenario design, media simulation, and use of video to support simulation. 23 minutes. Source: BP International, (216) 586-8291.

CAER: Planning Emergency Exercises. Explains how to plan and conduct four types of community emergency exercises. Source: Chemical Manufacturers Association, Washington, DC. Length: 10 minutes.

CBS News Report: UF-6 Release in Gore, Oklahoma. Example of how poor emergency response planning can hurt an organization; in this case, a Kerr-McGee plant was involved. 4:40 minutes. Source: CBS News.

Environmental Laws and Regulations: An Overview. Presents an overview of major environmental laws, including Clean Air Act, FIFRA, RCRA, Federal Water Pollution Control Act and Superfund/CERCLA. 16 minutes. Date: 1989. Source: Government Institutes Catalog, (301) 921-2355.

Fighting Terrorism: A National Security View. Command post-type exercise by well-known figures representing the members of the National Security Council. Involves limited outside contact in that they can deliver and receive telephone calls and written messages to their "aides." 58 minutes. Source: PBS Video, Alexandria, VA.

Incident Response: The First Responder's Role. Presents incident response procedures for first responders at the awareness and operations levels. 12 minutes. Date: 1992. Produced by Air & Water Technologies Corporation, available through Government Institutes Catalog, (301) 921-2355.

A Matter of When. A mass casualty incident simulation drill. The incident is a school fire and involves coordination of emergency medical personnel, firefighters, air rescue teams, hospital personnel, and specialty burn units. 11 minutes. Date: 1990. Source: National Burn Victim Foundation.

Miamisburg: Anatomy of a Response. Documentary of how exercises can aid in the preparation of response to an actual emergency. 19 minutes. Date: 1986. Source: Chemical Manufacturers Association, Washington, DC.

Media & Society Series: In the Face of Terrorism. Series of four tabletop exercises in which law professor Fred Friendly facilitates discussion by well-known media figures of the role and impact the media has on terrorism and terrorist situations. Good example of a facilitated tabletop exercise. Date: 1986.

Ready '88. Johnson Wax exercise in Racine County, Wisconsin. Includes brief display of moulage techniques. Very realistic emergency exercise, involving LEPCs. Also describes lessons learned. 40 minutes. Source: Quark Management Service, Racine, WI, (414) 632-5751.

Understanding Title III: Emergency Planning and Community Right-to-Know. Covers government, local, and business responsibilities; including emergency response procedures and interacting with community officials. 15 minutes. Date: 1987. Source: Government Institutes Catalog, (301) 921-2355.

When Disaster Strikes.... Documentary of a Chemolite exercise, involving the Racine (WI) County Local Emergency Planning Committee testing chemical emergency plans. 19 minutes. Date: 1988. Source: Quark Management Service, (414) 632-5751.

Index

E